**Disturbances in the linear model, estimatio**

# Disturbances in the linear model, estimation and hypothesis testing

**C. Dubbelman**
Econometric Institute,
Erasmus University Rotterdam.

*Martinus Nijhoff Social Sciences Division*
*Leiden | Boston 1978*

ISBN 90 207 0772 8



Printed in the Netherlands by Intercontinental Graphics

# Contents

# List of symbols

(see also Section 1.A)

# 1. Introduction

## 1.1. The general linear model

*All econometric research is based on a set of numerical data relating to certain economic quantities, and makes inferences from the data about the ways in which these quantities are related (Malinvaud 1970, p. 3).*

The linear relation is frequently encountered in applied econometrics. Let $y$ and $x$ denote two economic quantities, then the linear relation between $y$ and $x$ is formalized by:

$$y = \beta_1 + \beta_2 x$$

where $\beta_1$ and $\beta_2$ are constants. When $\beta_1$ and $\beta_2$ are known numbers, the value of $y$ can be calculated for every given value of $x$. Here $y$ is the dependent variable and $x$ is the explanatory variable.

In practical situations $\beta_1$ and $\beta_2$ are unknown. We assume that a set of $n$ observations on $y$ and $x$ is available. When plotting the observed pairs $(x_1, y_1), (x_2, y_2), \ldots, (x_n, y_n)$ into a diagram with $x$ measured along the horizontal axis and $y$ along the vertical axis it rarely occurs that all points lie on a straight line. Generally, no $b_1$ and $b_2$ exist such that $y_i = b_1 + b_2 x_i$ for $i = 1, 2, \ldots, n$. Unless the diagram clearly suggests another type of relation, for instance quadratic or exponential, it is customary to adopt linearity in order to keep the analysis as simple as possible. It is frequently possible to treat a nonlinear relation between $y$ and $x$ as a linear relation between new variables, which are transformations of $y$ and $x$.

The distance $v_i$ between the point $(x_i, y_i)$ of the scatter diagram and the point $(x_i, b_1 + b_2 x_i)$ on a straight line through the scatter is often called the error, $v_i = y_i - b_1 - b_2 x_i$, or:

$$y_i = b_1 + b_2 x_i + v_i$$

The word *error* is misleading in that it suggests that the distances would vanish if the observations did not suffer from measurement errors. Typically, the value of an economic quantity results from a complicated process in which numerous variables play a more or less important role. Therefore it is necessary to include a disturbance $u$ in the relation:

$$y = \beta_1 + \beta_2 x + u$$

The disturbance term accounts for the divergence between the assumed linear relationship and the true relationship between $y$ and $x$, and for the total, cumulated effect of all variables which are disregarded in the analysis at hand.

Essentially, a disturbance is to be regarded as a random variable, whose value cannot be observed. This implies that $y$, being a function of $u$, is itself a random variable. The linear relation can easily be extended to include more, say $k$, explanatory variables. Let $x_{ij}$ be the $i$th observation on the $j$th explanatory variable $x_j$, then the linear relation reads:

$$y = \sum_{j=1}^{k} \beta_j x_j + u$$

and the distance $v_i$ between a point $(x_{i1}, x_{i2}, \ldots, x_{ik}, y_i)$ and a line $y = \sum_{j=1}^{k} b_j x_j$ is defined as $v_i = y_i - \sum_{j=1}^{k} b_j x_{ij}$, or, in matrix notation:

$$\mathbf{y} = \mathbf{X}\mathbf{b} + \mathbf{v}$$

where:

$$\mathbf{y} = \begin{bmatrix} y_1 \\ y_2 \\ \vdots \\ y_n \end{bmatrix}; \quad \mathbf{X} = \begin{bmatrix} x_{11} & x_{12} & \ldots & x_{1k} \\ x_{21} & x_{22} & \ldots & x_{2k} \\ \vdots & & & \vdots \\ x_{n1} & x_{n2} & \ldots & x_{nk} \end{bmatrix}; \mathbf{b} = \begin{bmatrix} b_1 \\ b_2 \\ \vdots \\ b_k \end{bmatrix}; \quad \mathbf{v} = \begin{bmatrix} v_1 \\ v_2 \\ \vdots \\ v_n \end{bmatrix}$$

Here a constant term, like $\beta_1$ in the first relation, can be defined by $x_{i1} = 1$ for $i = 1, 2, \ldots, n$.

The introduction of random disturbances requires that assumptions be made concerning their probability distribution. Otherwise one may take $\beta_j = b_j$ with $b_j$ arbitrary, and the data can be used only to calculate outcomes of the disturbances. Other values of the $\beta$'s would yield another set of outcomes, and there is no criterion to decide which one of the two sets of outcomes is more plausible.

In this study we deal with the linear relation which together with a set of assumptions is called the general linear model. The model reads:

$$\mathbf{y} = \mathbf{X}\boldsymbol{\beta} + \mathbf{u} \tag{1.1}$$

where $\mathbf{y}$ is an $n$-element vector of observations on the dependent variable; $\mathbf{X}$ is an $n \times k$ matrix whose $j$th column consists of $n$ observations on the $j$th explanatory variable, $j = 1, 2, \ldots, k$; $\boldsymbol{\beta}$ is a $k$-element vector of unknown constants; $\mathbf{u}$ is an $n$-element vector of nonobservable outcomes of random variables. The matrix $\mathbf{X}$ is regarded as fixed, with rank $k$. The disturbance vector $u$ has the properties:

$$\mathcal{E}(\mathbf{u}) = \mathbf{o}, \qquad \mathcal{V}(\mathbf{u}) = \mathcal{E}(\mathbf{u}\mathbf{u}') = \sigma^2\boldsymbol{\Gamma}$$

where $\mathcal{E}(\mathbf{u})$ stands for the mathematical expectation of $\mathbf{u}$, $\mathcal{V}(\mathbf{u})$ stands for the covariance matrix of $\mathbf{u}$, $\sigma$ is a positive scalar constant, and $\boldsymbol{\Gamma}$ is an $n \times n$ positive definite symmetric matrix. It is assumed that the disturbances have a normal distribution, in brief, $\mathbf{u} \sim \mathcal{N}(\mathbf{o}, \sigma^2\boldsymbol{\Gamma})$. A prime denotes matrix transposition (see the appendix to this chapter).

The assumption of normality is not necessary for all types of analysis within the context of the linear model. Therefore this assumption is not always regarded as an element of the linear model. Normality is usually justified by an appeal to the central limit theorem.

For the sake of convenience, $\boldsymbol{\Gamma}$ is assumed to be nonsingular. In view of the transformation $(\mathbf{y} - \mathbf{c}) = \mathbf{X}\boldsymbol{\beta} + (\mathbf{u} - \mathbf{c})$ if the mean of $\mathbf{u}$ is known to be equal to $\mathbf{c}$, the assumption of zero mean does not cause any loss of generality. The assumption about the rank of $\mathbf{X}$ assures the existence of $(\mathbf{X}'\mathbf{A}\mathbf{X})^{-1}$ for arbitrary nonsingular $n \times n$

matrix **A**. To regard the matrix **X** as fixed means either that each of the explanatory variables (regressors) is regarded as nonstochastic, or that the regressors are stochastic but that the regressors and the disturbances are stochastically independent. The first case implies an odd treatment of the dependent variable and the regressors: the character of the variables is often similar, a regressor in one model is often the dependent variable in another model. In the second case, the model is called the regression model. This model specifies the conditional distribution of the dependent variable for each set of fixed values of the regressors.

In the present study we accept the general linear model. For a detailed discussion of the assumptions the reader is referred to Malinvaud (1970).

An important problem in applications of the general linear model is the estimation of the unknown $\boldsymbol{\beta}$, the vector of regression coefficients. The *best linear unbiased (BLU)* estimator is known to be equal to:

$$\hat{\boldsymbol{\beta}} = (\mathbf{X}'\Gamma^{-1}\mathbf{X})^{-1}\mathbf{X}'\Gamma^{-1}\mathbf{y} \tag{1.2}$$

with properties:

$$\mathcal{E}(\hat{\boldsymbol{\beta}}) = \boldsymbol{\beta}, \qquad \mathcal{V}(\hat{\boldsymbol{\beta}}) = \sigma^2 (\mathbf{X}'\Gamma^{-1}\mathbf{X})^{-1} \tag{1.3}$$

The estimator $\hat{\boldsymbol{\beta}}$ is a function of $\Gamma$, which matrix is specified by hypothesis. If the true covariance matrix of $\mathbf{u}$ differs from the hypothesized covariance matrix, then $\hat{\boldsymbol{\beta}}$ is not *best*.

The hypothesis $\Gamma = \mathbf{I}$ is popular, at least partly because it simplifies the calculations involved. When the validity of this specification is doubtful, a new estimate may be calculated on the basis of an alternative hypothesis. If the estimates differ significantly a choice must be made. The theory of hypothesis testing tries to formulate probability statements which may be helpful with respect to the choice problem. Most of the current testing procedures are unpopular, however, because of the amount of computation involved or because of the weakness of the probability statements.

In Chapter 2 we consider most powerful (however weak that may be) testing procedures. The test statistics, based on the vector of regression residuals $\mathbf{u}^* = \mathbf{y} - \mathbf{X}\hat{\boldsymbol{\beta}}$, have probability distributions which vary from **X** to **X**. When replacing $\mathbf{u}^*$ (the *BLU* disturbance estimator) by another disturbance estimator, it is possible to obtain test statistics with probability distributions which do not vary from **X** to **X**. This replacement causes some loss of power, generally, and it greatly reduces the amount of computational work. In Chapter 3 the *BLUF* (*F: fixed covariance matrix*) disturbance estimator is derived for an arbitrarily fixed covariance matrix. The specification of this covariance matrix is the subject of Chapter 4. In Chapter 5 the usefulness of several disturbance estimators with various covariance matrices is investigated by comparing powers of tests against positive autocorrelation and heterovariance in economic time series models with a small number of observations ($n \leqslant 20$).

The remainder of Chapter 1 contains a further introduction to the object of study. A survey of the relevant notions and notations in linear algebra is presented in the appendix to Chapter 1, together with some useful results.

## 1.2. *BLU* $\boldsymbol{\beta}$-estimation

Suppose two persons are asked to fit a straight line through the scatter diagram described at the beginning of this chapter. The two lines will almost certainly not coincide if they are drawn with the naked eye. Which one of the two lines is "best" can only be determined by means of a criterion.

One criterion is that the sum of the squared distances is a minimum, the so-called *least-squares* principle. The vector of distances is $\mathbf{v} = \mathbf{y} - \mathbf{X}\mathbf{b}$, and the least-squares principle implies to choose $\mathbf{b}$ such that $\mathbf{v}'\mathbf{v}$ be minimal. This $\mathbf{b}$ is found from putting the derivative of $\mathbf{v}'\mathbf{v} = \mathbf{y}'\mathbf{y} - 2\mathbf{y}'\mathbf{X}\mathbf{b} + \mathbf{b}'\mathbf{X}'\mathbf{X}\mathbf{b}$ with respect to $\mathbf{b}$ equal to zero, $-2\mathbf{X}'\mathbf{y} + 2\mathbf{X}'\mathbf{X}\mathbf{b} = \mathbf{o}$, so that $\mathbf{b} = (\mathbf{X}'\mathbf{X})^{-1}\mathbf{X}'\mathbf{y}$. The second-order derivative with respect to $\mathbf{b}$ is $2\mathbf{X}'\mathbf{X}$, which matrix is positive definite if the rank of $\mathbf{X}$ is $k$. Hence, $\mathbf{v}'\mathbf{v}$ is indeed minimal.

The above criterion attaches equal weights to all squared distances. Suppose that the distances are not equally important. Let $\mathbf{S}^{-1}\mathbf{v}$ be the vector of standardized distances, $\mathbf{S}^{-1}\mathbf{v} = \bar{\mathbf{y}} - \bar{\mathbf{X}}\mathbf{b}$, where $\bar{\mathbf{y}} = \mathbf{S}^{-1}\mathbf{y}$ and $\bar{\mathbf{X}} = \mathbf{S}^{-1}\mathbf{X}$. That is, the elements of $\mathbf{S}^{-1}\mathbf{v}$ are regarded as equally important. Minimization of the sum of squared standardized distances yields $\mathbf{b} = (\bar{\mathbf{X}}'\bar{\mathbf{X}})^{-1}\bar{\mathbf{X}}'\bar{\mathbf{y}}$. This vector $\mathbf{b}$ is called the *generalized least-squares (g. l. s.)* $\boldsymbol{\beta}$-estimator, whereas the same expression without bars is the *simple* or *ordinary least-squares (o. l. s.)* $\boldsymbol{\beta}$-estimator. For these b's the vectors $\mathbf{v} = \mathbf{y} - \mathbf{Xb}$ are called the *g. l. s.* and *o. l. s. regression residual* vectors.

Another criterion is the *BLU* principle. Here it is assumed that $\mathbf{y}$ is a vector of random variables with $\mathcal{E}(\mathbf{y}) = \mathbf{X}\boldsymbol{\beta}$ and $\mathcal{V}(\mathbf{y}) = \sigma^2\Gamma$. This implies that an estimator $\mathbf{b}$, which is a function of $\mathbf{y}$, is also a random variable. The *BLU* estimator $\hat{\boldsymbol{\beta}}$ of $\boldsymbol{\beta}$ is the *best* (*B*) of all estimators $\mathbf{b}$ which are *linear* (*L*) functions of $\mathbf{y}$ and which are *unbiased* (*U*), i.e. $\mathcal{E}(\mathbf{b}) = \boldsymbol{\beta}$, best in the sense that:

$$\mathcal{E}[(\mathbf{b} - \boldsymbol{\beta})'\mathbf{W}(\mathbf{b} - \boldsymbol{\beta})]$$

is minimal for $\mathbf{b} = \hat{\boldsymbol{\beta}}$, where $\mathbf{W}$ is an arbitrary positive definite symmetric $k \times k$ matrix; in brief, $\mathbf{W} = \mathbf{W}'_{(k)} > 0$. An alternative interpretation of the same best criterion is to say that:

$$\mathit{var}\ (\mathbf{g}'\mathbf{b})$$

is minimal for $\mathbf{b} = \hat{\boldsymbol{\beta}}$, where $\mathbf{g}$ is an arbitrary $k$-element vector. Below we show that $\hat{\boldsymbol{\beta}}$ is given by (1.2).

It is important to realize that the *g. l. s.* vector $\mathbf{b}$ with $\mathbf{S}$ such that $\mathbf{SS}' = \Gamma$ is equal to $\hat{\boldsymbol{\beta}}$, apart from their interpretation. Under the assumption that $\mathbf{y}$ has a normal distribution, still another criterion yields the same estimator $\hat{\boldsymbol{\beta}}$, namely the *maximum likelihood* principle (see e.g. Kmenta 1971, pp. 505-506). In the latter approach, linearity and unbiasedness are derived properties of $\hat{\boldsymbol{\beta}}$, whereas they are imposed in the *BLU* approach. Also, $\hat{\boldsymbol{\beta}}$ emerges as a *best unbiased* estimator if normality is assumed, best in the sense of smallest covariance matrix (see e.g. Koerts and Abrahamse 1969, pp. 16-18).

The coincidence of several criteria yielding the same estimator of $\boldsymbol{\beta}$ strongly justifies the wide use of $\hat{\boldsymbol{\beta}}$ in practical applications. An important feature of $\hat{\boldsymbol{\beta}}$ is that it depends on $\Gamma$. In cases where the specification of $\Gamma$ is uncertain, one may try to learn from the sample, the observation vector $\mathbf{y}$. Obviously, it is impossible to estimate all $\frac{1}{2}n(n + 1)$ different elements (not $n^2$, because of the symmetry) of $\Gamma$. In order to reduce the number of unknown elements of $\Gamma$, it is customary to assume that the elements are known functions of a small number of parameters (see Section 1.4). Then one is left with the problem of making inferences about these parameters. This is usually done on the basis of *o.l.s.* regression residuals (see Section 1.3). Section 1.5 contains the results of such parameter estimation from simulated first-order autocorrelated disturbances. Point estimates from small samples (we consider $n = 16$) appear to be very unreliable. Such estimates should be judged in the light of their probability distributions. This is what the theory of hypothesis testing does.

Any *linear unbiased (LU)* $\boldsymbol{\beta}$-estimator $\mathbf{b}$ can be written as $\mathbf{b} = \mathbf{A}'\mathbf{y}$ where $\mathbf{A}$ is an $n \times k$ matrix, independent of $\mathbf{y}$ and satisfying $\mathbf{A}'\mathbf{X} = \mathbf{I}_{(k)}$. This can be seen as follows. A linear function of $\mathbf{y}$ has the form $\mathbf{A}'\mathbf{y} + \mathbf{a}$, where both $\mathbf{A}$ and $\mathbf{a}$ are independent of $\mathbf{y}$. When linearity is interpreted in a strict sense, then $\mathbf{a} = \mathbf{o}$ (see e.g. Malinvaud 1970, p. 91). The function $\mathbf{b}$ is unbiased if $\mathcal{E}(\mathbf{b}) = \boldsymbol{\beta}$. Thus, we must have $\boldsymbol{\beta} = \mathcal{E}(\mathbf{A}'\mathbf{y} + \mathbf{a}) = \mathcal{E}(\mathbf{A}'\mathbf{X}\boldsymbol{\beta} + \mathbf{A}'\mathbf{u} + \mathbf{a}) = \mathbf{A}'\mathbf{X}\boldsymbol{\beta} + \mathbf{a}$, which equation must be an identity with respect to $\boldsymbol{\beta}$ since $\boldsymbol{\beta}$ is unknown. Hence, $\mathbf{a} = \mathbf{o}$ and $\mathbf{A}'\mathbf{X} = \mathbf{I}_{(k)}$. From the class of *LU* $\boldsymbol{\beta}$-estimators $\mathbf{b}$ we wish to find $\mathbf{b}$ such that $var(\mathbf{g}'\mathbf{b})$ be minimal for every $k$-element vector $\mathbf{g}$. That is, the variance of every weighted sum of the elements of $\mathbf{b}$ be minimal. Using $\mathbf{b} - \boldsymbol{\beta} = \mathbf{A}'\mathbf{u}$, we have:

$$var(\mathbf{g}'\mathbf{b}) = var(\mathbf{g}'\mathbf{A}'\mathbf{u}) = \mathcal{E}(\mathbf{g}'\mathbf{A}'\mathbf{u}\mathbf{u}'\mathbf{A}\mathbf{g}) = \sigma^2\mathbf{g}'\mathbf{A}'\boldsymbol{\Gamma}\mathbf{A}\mathbf{g} =$$

$$\sigma^2 tr(\mathbf{A}\mathbf{g}\mathbf{g}'\mathbf{A}'\boldsymbol{\Gamma})$$

Let $\mathbf{g}_i$ be the $i$th column of an arbitrary $k \times l$ matrix $\mathbf{G}$ with rank $l$, then:

$$\sum_{i=1}^{l} var(\mathbf{g}_i'\mathbf{b}) = \sum_{i=1}^{l} \sigma^2 tr(\mathbf{A}\mathbf{g}_i\mathbf{g}_i'\mathbf{A}'\boldsymbol{\Gamma}) = \sigma^2 tr(\mathbf{A}\mathbf{G}\mathbf{G}'\mathbf{A}'\boldsymbol{\Gamma})$$

Momentarily, one may think of $l = 1$. We wish to minimize $tr(\mathbf{A}\mathbf{G}\mathbf{G}'\mathbf{A}'\boldsymbol{\Gamma})$ with respect to $\mathbf{A}$, subject to $\mathbf{A}'\mathbf{X} = \mathbf{I}$. Note that:

$$\mathcal{E}[(\mathbf{b} - \boldsymbol{\beta})'\mathbf{W}(\mathbf{b} - \boldsymbol{\beta})] = \mathcal{E}[\mathbf{u}'\mathbf{A}\mathbf{W}\mathbf{A}'\mathbf{u}] = \sigma^2 tr(\mathbf{A}\mathbf{W}\mathbf{A}'\boldsymbol{\Gamma})$$

so that the two best criteria coincide if $l = k$. The minimization problem is usually solved by means of the Lagrange technique. This technique says: differentiate:

$$tr(\mathbf{AGG'A'\Gamma}) - tr[\mathbf{L'(A'X - I)}]$$

with respect to both **A** and **L**, where **L** is the matrix of Lagrange multipliers, and put the derivatives[1] equal to **0**. Doing this, we find the Lagrange conditions:

$$2\mathbf{\Gamma AGG'} = \mathbf{XL'} \text{ and } \mathbf{A'X} = \mathbf{I}$$

It follows that $\mathbf{L}' = 2(\mathbf{X}'\mathbf{\Gamma}^{-1}\mathbf{X})^{-1}\mathbf{GG}'$, so that we have:

$$\mathbf{AG} = \mathbf{\Gamma}^{-1}\mathbf{X}(\mathbf{X}'\mathbf{\Gamma}^{-1}\mathbf{X})^{-1}\mathbf{G} \text{ and } \mathbf{A'X} = \mathbf{I}$$

Hence, $\mathbf{A} = \mathbf{A}_0 + \mathbf{F}$, where $\mathbf{A}_0 = \mathbf{\Gamma}^{-1}\mathbf{X}(\mathbf{X}'\mathbf{\Gamma}^{-1}\mathbf{X})^{-1}$ and **F** is an $n \times k$ matrix satisfying $\mathbf{FG} = \mathbf{0}$ and $\mathbf{F'X} = \mathbf{0}$. The restriction $\mathbf{F'X} = \mathbf{0}$ implies that all columns of **F** are vectors lying in $\mathcal{M}(\mathbf{X})^{\perp} = \mathcal{M}(\mathbf{M})$, where $\mathbf{M} = \mathbf{I} - \mathbf{X}(\mathbf{X}'\mathbf{X})^{-1}\mathbf{X}'$ (see Sections 1.A.6.1 and 1.A.6.2). Hence, each matri[x] **F** satisfying $\mathbf{F'X} = \mathbf{0}$ can be written as $[\mathbf{I} - \mathbf{X}(\mathbf{X}'\mathbf{X})^{-1}\mathbf{X}']\mathbf{C}_1$ for some matrix $\mathbf{C}_1$ with the same order as the order of **F**. Analogously, $\mathbf{FG} = \mathbf{0}$ implies $\mathbf{F} = \mathbf{C}[\mathbf{I} - \mathbf{G}(\mathbf{G}'\mathbf{G})^{-1}\mathbf{G}']$ for some matrix **C**. Using $\mathbf{M} = \mathbf{M}^2$, we have $\mathbf{F} = \mathbf{MC}_1 = \mathbf{M}^2\mathbf{C}_1 = \mathbf{MF} = \mathbf{MC}[\mathbf{I} - \mathbf{G}(\mathbf{G}'\mathbf{G})^{-1}\mathbf{G}']$. Therefore the Lagrange conditions are satisfied for:

$$\mathbf{A} = \mathbf{A}_0 + \mathbf{MC}[\mathbf{I} - \mathbf{G}(\mathbf{G}'\mathbf{G})^{-1}\mathbf{G}']$$

where **C** is an arbitrary $n \times k$ matrix. Clearly, $\mathbf{A} = \mathbf{A}_0$ is the unique solution if $l = k$. For instance, if one wishes to minimize the variances of the elements of **b** individually, then $\mathbf{G} = \mathbf{I}_{(k)}$ and $\mathbf{b} = \hat{\boldsymbol{\beta}}$. In particular, all elements of the vector $\hat{\mathbf{y}} = \mathbf{Xb}$ are weighted sums of the elements of **b**. The variance of every element of $\hat{\mathbf{y}}$ is minimized by $\mathbf{b} = \hat{\boldsymbol{\beta}}$. The fact that $\mathbf{A} = \mathbf{A}_0$ indeed minimizes $tr(\mathbf{AGG'A'\Gamma})$ is usually proved by comparing the values of the trace for $\mathbf{A} = \mathbf{A}_0$ and $\mathbf{A} = \mathbf{A}_1 = \mathbf{A}_0 + \mathbf{V}$, where **V** is a nonzero $n \times k$ matrix. Obviously, **V** satisfies $\mathbf{V'X} = \mathbf{0}$ in order that $\mathcal{E}(\mathbf{A}_1'\mathbf{y}) = \boldsymbol{\beta}$. Using $\mathbf{V}'\mathbf{\Gamma A}_0 = \mathbf{V}'\mathbf{X}(\mathbf{X}'\mathbf{\Gamma}^{-1}\mathbf{X})^{-1} = \mathbf{0}$ it is found that $tr(\mathbf{A}_1\mathbf{GG}'\mathbf{A}_1'\mathbf{\Gamma}) - tr(\mathbf{A}_0\mathbf{GG}'\mathbf{A}_0'\mathbf{\Gamma}) = tr(\mathbf{VGG'V'\Gamma}) = tr[(\mathbf{\Gamma}^{1/2}\mathbf{VG})(\mathbf{\Gamma}^{1/2}\mathbf{VG})']$. The latter trace is the sum of the squared elements of the matrix $\mathbf{\Gamma}^{1/2}\mathbf{VG}$, which is nonnegative. It is strictly positive when $l = k$.

1. Generally, $\dfrac{\partial}{\partial \mathbf{A}}[tr(\mathbf{AB})] = \mathbf{B}'$ and $\dfrac{\partial}{\partial \mathbf{A}}[tr(\mathbf{ABA'C})] = \mathbf{CAB} + \mathbf{C'AB'}$.

## 1.3 *BLU* disturbance estimation

A vector **v** which can be written as $\mathbf{v} = \mathbf{y} - \mathbf{Xb}$ is called a regression residual vector. When **b** is a *LU* $\boldsymbol{\beta}$-estimator, so that $\mathbf{b} = \mathbf{A}'\mathbf{y}$ with $\mathbf{A}'\mathbf{X} = \mathbf{I}$, we have $\mathbf{v} = (\mathbf{I} - \mathbf{XA}')\mathbf{y}$, where $\mathbf{I} - \mathbf{XA}'$ is idempotent with rank $n - k$. The elements of $\mathbf{u}^* = \mathbf{y} - \mathbf{X}\hat{\boldsymbol{\beta}}$ are the *g.l.s.* regression residuals:

$$\mathbf{u}^* = \mathbf{y} - \mathbf{X}\hat{\boldsymbol{\beta}} = \mathbf{M}^*\mathbf{y}$$

where:

$$\mathbf{M}^* = \mathbf{I} - \mathbf{X}(\mathbf{X}'\boldsymbol{\Gamma}^{-1}\mathbf{X})^{-1}\mathbf{X}'\boldsymbol{\Gamma}^{-1}$$

The vector $\mathbf{u}^*$ is that vector **v** which minimizes $\mathscr{E}[(\mathbf{v} - \mathbf{u})'\mathbf{Q}(\mathbf{v} - \mathbf{u})]$ subject to $\mathbf{v} = \mathbf{B}'\mathbf{y}$ with **B** independent of **y** and $\mathbf{B}'\mathbf{X} = \mathbf{0}$, where **Q** is an arbitrary matrix satisfying $\mathbf{Q} = \mathbf{Q}'_{(n)} > 0$. The derivation is analogous to the derivation of the *BLU* $\boldsymbol{\beta}$-estimator in the previous section. Because of the analogy, $\mathbf{u}^*$ is called the *BLU* disturbance estimator.[2] Note that the meaning of the words estimator and unbiased has been changed. In the case of $\boldsymbol{\beta}$, the elements to be estimated are unknown constants, whereas they are random variables in the case of **u**. In the former case, $\mathscr{E}(\mathbf{b} - \boldsymbol{\beta}) = \mathbf{o}$ and $\mathscr{E}(\mathbf{b}) = \boldsymbol{\beta}$ are equivalent expressions of the unbiasedness of **b**; in the latter case, $\mathscr{E}(\mathbf{v} - \mathbf{u}) = \mathbf{o}$ is equivalent to $\mathscr{E}(\mathbf{v}) = \mathscr{E}(\mathbf{u}) = \mathbf{o}$. With these interpretations in mind, it is easily verified that any *LU* disturbance estimator can be written as $\mathbf{v} = \mathbf{B}'\mathbf{y}$ with **B** independent of **y** and $\mathbf{B}'\mathbf{X} = \mathbf{0}$.

In the general model (1.1) it is assumed that $\boldsymbol{\Gamma}$ satisfies $\boldsymbol{\Gamma} = \boldsymbol{\Gamma}'_{(n)} > 0$. Then $\boldsymbol{\Gamma}$ can be written as $\boldsymbol{\Gamma} = \mathbf{SS}'$ with **S** nonsingular. For instance, $\mathbf{S} = \boldsymbol{\Gamma}^{1/2}$. Note that **S** is not unique (see Result 1.A.6.3). When $\boldsymbol{\Gamma} = \mathbf{I}$, the model is called the simple model; we may take $\mathbf{S} = \mathbf{I}$. We define:

$$\bar{\mathbf{y}} = \mathbf{S}^{-1}\mathbf{y} \qquad \bar{\mathbf{X}} = \mathbf{S}^{-1}\mathbf{X} \qquad \bar{\mathbf{u}} = \mathbf{S}^{-1}\mathbf{u}$$

$$\mathbf{M} = \mathbf{I}_{(n)} - \mathbf{X}(\mathbf{X}'\mathbf{X})^{-1}\mathbf{X}' \qquad \bar{\mathbf{M}} = \mathbf{I}_{(n)} - \bar{\mathbf{X}}(\bar{\mathbf{X}}'\bar{\mathbf{X}})^{-1}\bar{\mathbf{X}}'$$

$$\hat{\mathbf{u}} = \mathbf{My} = \mathbf{Mu} \qquad \hat{\bar{\mathbf{u}}} = \bar{\mathbf{M}}\bar{\mathbf{y}} = \bar{\mathbf{M}}\bar{\mathbf{u}}$$

2. Preference is sometimes given to predictor or approximator instead of estimator (see e.g. Ramsey 1969).

Note that $\mathbf{My} = \mathbf{Mu}$ follows from $\mathbf{MX} = \mathbf{0}$; analogously, $\overline{\mathbf{M}}\overline{\mathbf{y}} = \overline{\mathbf{M}}(\overline{\mathbf{X}}\boldsymbol{\beta} + \overline{\mathbf{u}}) = \mathbf{0}\boldsymbol{\beta} + \overline{\mathbf{M}}\overline{\mathbf{u}} = \overline{\mathbf{M}}\overline{\mathbf{u}}$. It is easily verified that $\hat{\overline{\mathbf{u}}} = \mathbf{S}^{-1}\mathbf{u}^*$ and that $\mathbf{M}^*\mathbf{S} = \mathbf{S}\overline{\mathbf{M}}$. Further:

$$\mathbf{M} = \mathbf{M}' = \mathbf{M}^2 \qquad rank(\mathbf{M}) = tr(\mathbf{M}) = n - k$$

$$\overline{\mathbf{M}} = \overline{\mathbf{M}}' = \overline{\mathbf{M}}^2 \qquad rank(\overline{\mathbf{M}}) = tr(\overline{\mathbf{M}}) = n - k$$

$$\mathbf{M}^* = \mathbf{M}^*\mathbf{M}^* \neq \mathbf{M}^{*\prime} \qquad rank(\mathbf{M}^*) = rank(\mathbf{M}^*\mathbf{S}) = rank(\mathbf{S}\overline{\mathbf{M}}) = rank(\overline{\mathbf{M}})$$

In accordance with Results 1.A.6.1 and 1.A.6.2:

$$\mathcal{M}(\mathbf{M})^{\perp} = \mathcal{M}[\mathbf{X}(\mathbf{X}'\mathbf{X})^{-1}\mathbf{X}'] = \mathcal{M}(\mathbf{X})$$

$$\mathcal{M}(\overline{\mathbf{M}})^{\perp} = \mathcal{M}(\overline{\mathbf{X}})$$

Since $\mathbf{Xb}$ is a vector in $\mathcal{M}(\mathbf{X})$, this space is called the regression space; while $\mathcal{M}(\mathbf{M})$ is called the error space, since $\hat{\mathbf{u}} = \mathbf{My}$ lies in it. These spaces can, of course, be spanned by orthonormal bases, the former consisting of $k$ vectors and the latter consisting of $n$-$k$ vectors. We define the $n \times k$ matrices $\mathbf{R}$ and $\overline{\mathbf{R}}$, and the $n \times (n-k)$ matrices $\mathbf{N}$ and $\overline{\mathbf{N}}$ as follows: the column vectors of $\mathbf{R}$ and $\overline{\mathbf{R}}$ form orthonormal bases of $\mathcal{M}(\mathbf{X})$ and $\mathcal{M}(\overline{\mathbf{X}})$, respectively; the column vectors of $\mathbf{N}$ and $\overline{\mathbf{N}}$ form orthonormal bases of $\mathcal{M}(\mathbf{M})$ and $\mathcal{M}(\overline{\mathbf{M}})$, respectively. Like $\mathbf{S}$, the matrices $\mathbf{R}$, $\overline{\mathbf{R}}$, $\mathbf{N}$ and $\overline{\mathbf{N}}$ are not unique. They satisfy:

$$[\mathbf{R} \vdots \mathbf{N}]' = [\mathbf{R} \vdots \mathbf{N}]^{-1} \qquad [\overline{\mathbf{R}} \vdots \overline{\mathbf{N}}]' = [\overline{\mathbf{R}} \vdots \overline{\mathbf{N}}]^{-1}$$

$$\mathbf{R}'\mathbf{R} = \overline{\mathbf{R}}'\overline{\mathbf{R}} = \mathbf{I}_{(k)}$$

$$\mathbf{N}'\mathbf{N} = \overline{\mathbf{N}}'\overline{\mathbf{N}} = \mathbf{I}_{(n-k)}$$

$$\mathbf{R}\mathbf{R}' + \mathbf{N}\mathbf{N}' = \overline{\mathbf{R}}\overline{\mathbf{R}}' + \overline{\mathbf{N}}\overline{\mathbf{N}}' = \mathbf{I}_{(n)}$$

$$\mathbf{R}'\mathbf{N} = \overline{\mathbf{R}}'\overline{\mathbf{N}} = \mathbf{0}$$

Since the columns of $\mathbf{R}$ form a basis of $\mathcal{M}(\mathbf{X})$, we have $\mathbf{X} = \mathbf{RG}$ for some nonsingular $k \times k$ matrix $\mathbf{G}$. It follows that:

$$\mathbf{M} = \mathbf{I} - \mathbf{RG}(\mathbf{G}'\mathbf{R}'\mathbf{RG})^{-1}\mathbf{G}'\mathbf{R}' = \mathbf{I} - \mathbf{RR}' = \mathbf{NN}'$$

Analogously,

$$\overline{\mathbf{M}} = \mathbf{I} - \overline{\mathbf{R}}\overline{\mathbf{R}}' = \overline{\mathbf{N}}\overline{\mathbf{N}}'$$

The columns of $\mathbf{N}$ (of $\overline{\mathbf{N}}$) are eigenvectors of $\mathbf{M}$ (of $\overline{\mathbf{M}}$) corresponding to its $n$-$k$ eigenvalues which are equal to 1; and the columns of $\mathbf{R}$ (of $\overline{\mathbf{R}}$) are eigenvectors of $\mathbf{M}$ (of $\overline{\mathbf{M}}$) corresponding to its $k$ zero eigenvalues. When writing $\mathbf{X} = \mathbf{RG}$, we may say that $\mathbf{R}$ represents the space spanned by the columns of $\mathbf{X}$, and $\mathbf{G}$ represents the location of $\mathbf{X}$ within that space. It is important to observe that $\mathbf{M}$ and $\hat{\mathbf{u}}$ do not depend on $\mathbf{G}$. For instance, when studying a statistic based on $\hat{\mathbf{u}}$, it is to be realized that, generally, the statistic and its probability distribution depend on $\mathbf{X}$, not on the location of its column vectors within the regression space, but only on the regression space itself, which can always be represented by an orthogonal matrix $\mathbf{R}$. We make use of this fact in Section 1.5, where an $\mathbf{X}$-matrix must be chosen for the simulation. We remark that $\overline{\mathbf{R}} \neq \mathbf{S}^{-1}\mathbf{R}$ and $\overline{\mathbf{N}} \neq \mathbf{S}^{-1}\mathbf{N}$, generally. For instance, $\overline{\mathbf{R}} = \mathbf{S}^{-1}\mathbf{R}$ would imply $\mathbf{R}'\mathbf{\Gamma}^{-1}\mathbf{R} = \mathbf{I}_{(k)}$, which establishes a relation between $\mathbf{X}$ and $\mathbf{\Gamma}$, while there is no such assumption in the general linear model.

## 1.4. Autocorrelation and heterovariance

The simplest hypothesis concerning the covariance matrix of the disturbances is that the disturbances are mutually uncorrelated and they have common variance. In brief, $\mathcal{V}(\mathbf{u}) = \sigma^2\mathbf{\Gamma} = \sigma^2\mathbf{I}$. In cross-section analysis, and to a lesser extent in time series analysis, the assumption of common variance (homoskedasticity) is often doubtful, i.e. different diagonal elements of $\mathbf{\Gamma}$ (heteroskedasticity, heterovariance) are plausible for one reason or another. In order to reduce the number of unknown elements [the $n$ variances $\sigma_1^2, \sigma_2^2, \ldots, \sigma_n^2$ on the diagonal of $\mathcal{V}(\mathbf{u})$], one frequently assumes that the variances are associated with some known variable. For instance, in a regression on family consumption, less variation in consumption may be expected for low-income families than for high-income families. In such a regression the variance for the $i$th family may be taken proportional to the level of family income $Y_i$ or a function of this level. A convenient relation would be:

$$\sigma_i^2 = \sigma^2 Y_i^{\delta} \qquad \text{(see Kmenta 1971, p. 258)}$$

This assumption specifies $\mathcal{V}(\mathbf{u})$ up to the parameters $\sigma^2$ and $\delta$. Note that $\delta = 0$ implies homoskedasticity. The assumption of no correlation is likely to be violated in many time series analyses. Most of the work in the context of dependence has been done on the assumption of the stationary stochastic (Markov) process defined by:

$$u_i = \rho u_{i-1} + \epsilon_i \qquad i = \ldots, 1, 0, 1, \ldots \tag{1.4}$$

with $|\rho| < 1$, where $\epsilon_i$ has a normal distribution with zero mean and variance $\sigma_\epsilon^2$ (independent of $i$), and $\epsilon_i$ is stochastically independent of $\epsilon_j$ and $u_j$ for $j = i-1, i-2, i-3, \ldots$ Then a vector $\mathbf{u}$ consisting of $n$ successive $u$-elements has a normal distribution with zero mean and covariance matrix $\left\{\sigma_\epsilon^2/(1-\rho^2)\right\}$ $\Gamma$ with:

$$\Gamma = \begin{bmatrix} 1 & \rho & \rho^2 & \rho^3 & \ldots & \rho^{n-1} \\ \rho & 1 & \rho & \rho^2 & \ldots & \rho^{n-2} \\ \rho^2 & \rho & 1 & \rho & \ldots & \rho^{n-3} \\ \vdots & & & & & \vdots \\ \rho^{n-1} & \rho^{n-2} & \rho^{n-3} & \rho^{n-4} & \ldots & 1 \end{bmatrix} \tag{1.5}$$

Disturbances having this distribution are said to be autocorrelated, with autocorrelation parameter $\rho$. The assumption specifies $\mathcal{V}(\mathbf{u})$ up to the parameters $\sigma^2 = \sigma_\epsilon^2/(1-\rho^2)$ and $\rho$. Note that $\rho = 0$ implies independence. Parameters like $\sigma^2$, $\delta$, and $\rho$ can be estimated. From:

$$\mathcal{E}(\mathbf{u}^{*\prime}\Gamma^{-1}\mathbf{u}^*) = \mathcal{E}[tr(\Gamma^{-1}\mathbf{u}^*\mathbf{u}^{*\prime})] = tr[\Gamma^{-1}\mathcal{E}(\mathbf{u}^*\mathbf{u}^{*\prime})] =$$
$$tr(\Gamma^{-1}\sigma^2\mathbf{M}^*\Gamma) = \sigma^2 tr(\mathbf{M}^*) = (n\text{-}k)\sigma^2$$

it follows, that when $\Gamma$ is known, $\mathbf{u}^{*\prime}\Gamma^{-1}\mathbf{u}^* / (n\text{-}k)$ is an unbiased estimator of $\sigma^2$. Estimation of $\delta$ and $\rho$ is much more complicated (see e.g. Kmenta 1971, pp. 257-264, 284-292). Though we are not really concerned with estimation of such parameters in this study, we investigate the properties of one of the current $\rho$-estimators in the next section. The only aim is to stress the need of hypothesis testing in this area.

## 1.5. Autocorrelation simulated and estimated

Suppose that a set of mutually stochastically independent standard normal random drawings $\{\epsilon_i\}$ is given. From these values one can calculate a vector **u**, whose elements are autocorrelated, with autocorrelation parameter $\rho$, in accordance with (1.4). In this case both **u** and $\rho$ are known. Given the vector **u**, one may try to estimate $\rho$. The estimate can be compared with the true value of $\rho$. This is done below. We take $n = 16$. A computer subroutine generates $\epsilon_0, \epsilon_1, \ldots, \epsilon_{16}$. The values $u_0, u_1, \ldots, u_{16}$ follow from:

$$u_0 = \epsilon_0/\sqrt{1-\rho^2}$$

$$u_i = \rho u_{i-1} + \epsilon_i \qquad i = 1, 2, \ldots, 16$$

Then $u_i$ has a normal distribution, $\mathcal{E}(u_i) = 0$, and $\mathcal{E}(u_i u_j) = \rho^{|i-j|}/(1-\rho^2)$ for $i = 0, 1, \ldots, 16$ and $j = 0, 1, \ldots, 16$. The *o.l.s.* estimator $\hat{\rho}$ of $\rho$ in:

$$u_i = \rho u_{i-1} + \epsilon_i \qquad i = 1, 2, \ldots, n$$

is:

$$\hat{\rho} = \sum_{i=1}^{n} u_i u_{i-1} \Big/ \sum_{i=1}^{n} u_{i-1}^2$$

For each of the values -0.9, -0.8, . . . , 0.9 for $\rho$ (the true values of $\rho$) we calculated $\hat{\rho}$, thus obtaining 19 estimates. The procedure was repeated: starting with a new set of values $\epsilon_0, \epsilon_1, \ldots, \epsilon_{16}$, and ending with another 19 estimates. After 1000 repetitions, we had 1000 estimates $\hat{\rho}$ when the true $\rho$ is equal to -0.9, another 1000 estimates $\hat{\rho}$ when the true $\rho$ is equal to -0.8, and so on. Given $\rho$, we calculated the mean $\bar{\hat{\rho}}$ and the variance $\nu$ of the 1000 estimates $\hat{\rho}$; $\bar{\hat{\rho}} = \Sigma\hat{\rho}/1000$ and $\nu = \Sigma(\hat{\rho} - \bar{\hat{\rho}})^2/1000$. The results are very well summarized by: given $\rho$, then $\bar{\hat{\rho}} = 0.9\rho$ and $\nu = (3-2\rho^2)/54$. Thus $\hat{\rho}$ underestimates $\rho$ by some 10 percent, while the standard error $\sqrt{\nu}$ varies from 0.16 (for $|\rho| = 0.9$) to 0.24 (for $\rho = 0.0$). The conclusion is that $\hat{\rho}$ is an unreliable estimator of $\rho$, because of the large variance. And the situation becomes even worse when we consider practical applications of the linear model, as follows.

Above we calculated $\hat{\rho}$ from the vector **u**. In the linear model, **u** is unknown. Only **y** and **X** can be observed. With respect to the estimation of $\rho$, several procedures have been proposed (see e.g. Kmenta 1971, pp. 284-289). Theil (1971, p. 254) proposed an estimator, which we denote by $\hat{\rho}_T$, based on the *o.l.s.* regression residual vector $\hat{\mathbf{u}} = [\hat{u}_1 \hat{u}_2 \ldots \hat{u}_n]'$:

$$\hat{\rho}_T = (n-k) \sum_{i=2}^{n} \hat{u}_i \hat{u}_{i-1} / (n-1), \sum_{i=1}^{n} \hat{u}_i^2$$

The formulas of $\hat{\rho}$ and $\hat{\rho}_T$ differ in three respects: in the latter formula $\hat{\mathbf{u}}$ is used instead of **u**, the multiplicative scalar (n-k)/($n$-1) is added, and $i$=1 is omitted in the summation in the numerator. The omission of $i$=1 is caused by the fact that $\hat{u}_0$ is not available. When the numerator in the formula of $\hat{\rho}$ would be replaced by

$n \sum_{i=2}^{n} u_i u_{i-1} / (n-1)$, which probably affects $\hat{\rho}$ very little, then the

difference between $\hat{\rho}$ and $n\hat{\rho}_T/(n-k)$ is the use of $\hat{\mathbf{u}}$ instead of **u**.

For both $\rho$ = 0.0 and $\rho$ = 0.8 (two true values of $\rho$) we generated 10000 vectors **u**, as indicated above. Figure 1.1 shows the frequency diagrams of the estimates $\hat{\rho}$ (from the first 1000 of $\hat{\rho}$ we calculated $\bar{\hat{\rho}}$ and $\nu$, see above). In order to trace the effect of dealing with $\hat{\mathbf{u}}$ instead of **u**, we selected a 16 × 3 **X**-matrix which can be regarded as typical of economic time series (namely $\mathbf{X} = [\mathbf{h}_1^* \vdots \mathbf{h}_2^* \vdots \mathbf{h}_3^*]$, see Chapter 4), and calculated the corresponding **M**-matrix. All vectors **u** were then premultiplied by this **M**, so as to obtain vectors $\hat{\mathbf{u}} = \mathbf{M}\mathbf{u}$. From these vectors we computed estimates $\hat{\rho}_T$, 10000 for $\rho$ = 0.0 and 10000 for $\rho$ = 0.8. The frequency diagrams of these estimates are presented in Figure 1.2. Regarding these as approximations of conditional probability density functions, the left-hand histogram with the condition $\rho$ = 0.0 and the right-hand with the condition $\rho$ = 0.8, we find an estimation bias of about -0.2 when $\rho$ = 0.0 and about -0.6 when $\rho$ = 0.8. Even if one were to adopt some bias correction (where it is to be realized that the bias depends on the specification of **M**), the variance of $\hat{\rho}_T$ is so large that the estimator, either corrected or uncorrected, remains very unreliable. A

*Figure 1.1.* Frequency diagram of 10000 $\hat{\rho}$'s for $\rho = 0.0$ and of 10000 $\hat{\rho}$'s for $\rho = 0.8$.

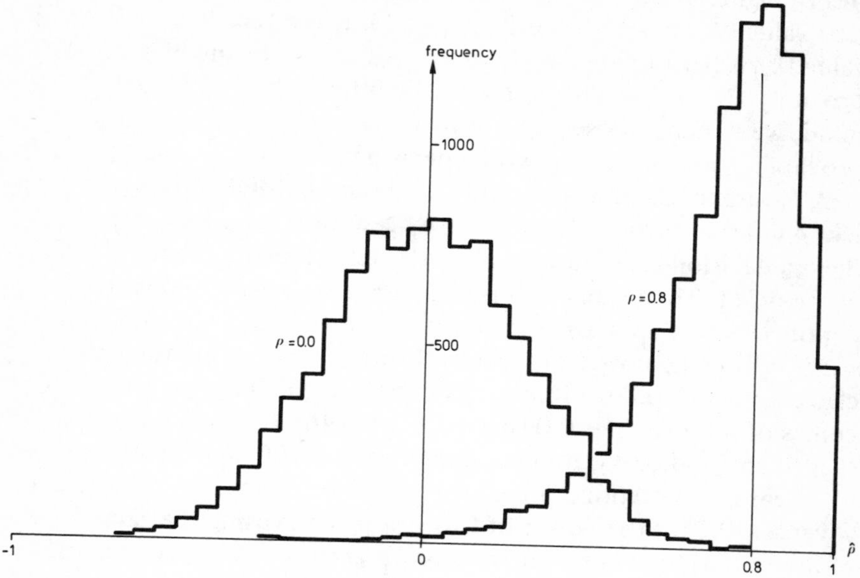


*Figure 1.2.* Frequency diagram of 10000 $\hat{\rho}_T$'s for $\rho = 0.0$ and of 10000 $\hat{\rho}_T$'s for $\rho = 0.8$.

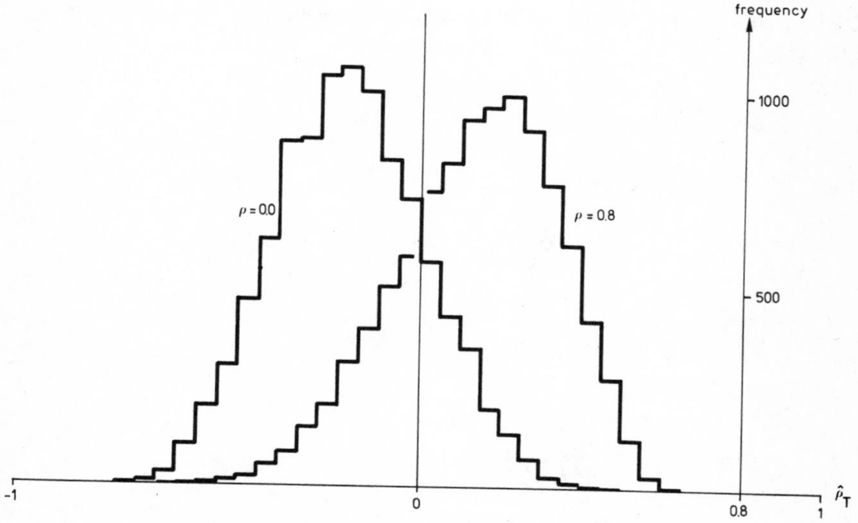

great part of the intervals of $\hat{\rho}_T$, where the two (approximate) densities in Figure 1.2 are positive, coincides. The question whether the true value of $\rho$ is equal to 0.0 or to 0.8 if it is given that $\hat{\rho}_T$ takes the value 0.2 is hard to answer. If we were unaware of the bias, we would probably choose $\rho = 0.0$. But with the information of Figure 1.2 in mind, we would choose $\rho = 0.8$. In all events, there is always a positive probability of making the wrong decision.

A natural decision rule would be: choose beforehand a value $s$, reject $\rho = 0.0$ in favour of $\rho = 0.8$ if $\hat{\rho}_T > s$, otherwise accept $\rho = 0.0$. Such a decision rule is a test, $\hat{\rho}_T$ is the test statistic, $s$ is the significance point. The probability that $\hat{\rho}_T$ exceeds $s$ if $\rho = 0.0$ is true, denoted by $Pr\,[\hat{\rho}_T > s | \rho = 0.0]$, is the significance level, and $Pr\,[\hat{\rho}_T > s | \rho = 0.8]$ is the power of the test. For instance, fixing the significance level at 0.05, we found from the table of frequencies of outcomes of $\hat{\rho}_T$ given $\rho = 0.0$ that $s = 0.145$ (466 outcomes are greater than 0.150 and 354 outcomes lie between 0.100 and 0.150, so that, by linear interpolation, 500 of the 10000 outcomes exceed 0.145). Given $s = 0.145$, the power of this test is 0.52 (using the same type of interpolation, 5211 outcomes of $\hat{\rho}_T$ given $\rho = 0.8$ exceed 0.145). An interesting question is whether it would be possible to realize a higher power at the same significance level, by means of another test statistic. The "practically best" (see the discussion in Section 2.1) test in this respect is the exact Durbin-Watson test, using the test statistic:

$$d = \sum_{i=2}^{n} (\hat{u}_i - \hat{u}_{i-1})^2 \Big/ \sum_{i=1}^{n} \hat{u}_i^2$$

From:

$$\sum_{i=2}^{n} (\hat{u}_i - \hat{u}_{i-1})^2 = \sum_{i=2}^{n} (\hat{u}_i^2 + \hat{u}_{i-1}^2) - 2 \sum_{i=2}^{n} \hat{u}_i \hat{u}_{i-1}$$
$$\approx 2\,\frac{n-1}{n} \sum_{i=1}^{n} \hat{u}_i^2 - 2 \sum_{i=2}^{n} \hat{u}_i \hat{u}_{i-1}$$

it is found that:

$$d \approx 2\left(\frac{n-1}{n} - \frac{n-1}{n-k}\hat{\rho}_T\right)$$

Hence a test in terms of $\hat{\rho}_T$ is almost equivalent to a test in terms of $d$. For instance, taking $n = 16$ and $k = 3$, the above test defined by

$\hat{\rho}_T > s = 0.145$ is equivalent to the test defined by

$2\,(\frac{15}{16} - \frac{15}{13}\,\hat{\rho}_T) < 2\,(\frac{15}{16} - \frac{15}{13}\,0.145) = 1.540$, and the power

remains 0.52. At the same significance level, the significance point of the exact Durbin-Watson test is 1.539 and the power is 0.543. Therefore the reliability of $\hat{\rho}_T$ (in the sense of distinguishing between $\rho = 0.0$ and $\rho = 0.8$) cannot be improved substantially: the information, provided by the sample, which is contained in $\hat{\rho}_T$ is almost equivalent to the information contained in $d$.

Summarizing, we can say that, in the first place, every estimator of $\rho$ leaves us with a great probability of making a wrong decision; in the second place, the relevant probability distributions should be taken into account when judging an outcome of an estimator; in the third place, a test quantifies (but not removes) the error probabilities; and in the fourth place, in the above example, it does not make sense to estimate $\rho$ by means of $\hat{\rho}_T$ if the Durbin-Watson test statistic has been calculated already, because of the close relationship between $d$ and $\hat{\rho}_T$.

## 1.A. Appendix

### *1.A.1. Notation of vectors and matrices*

Upper-case[1] boldface symbols denote matrices: $\mathbf{A}$, $\mathbf{\Gamma}$, $\mathbf{D}^*$, $\bar{\mathbf{X}}$; lower-case boldface symbols denote vectors: $\mathbf{b}$, $\mathbf{h}^*$, $\mathbf{e}_i$, $\boldsymbol{\lambda}$. A prime denotes matrix transposition. A lower-case boldface symbol without a prime always denotes a column vector, e.g. $\mathbf{a}$ is a column vector and $\mathbf{a}'$ is a row vector. The elements of an $n$-element vector $\mathbf{a}$ are denoted by $a_1, a_2, \ldots, a_n$. An $n \times k$ matrix $\mathbf{A}$ has $n$ rows and $k$ columns. The element $a_{ij}$ of $\mathbf{A}$ is the element on row $i$ and column $j$. If $n = k$, $\mathbf{A}$ is square of order $n$. If we wish to indicate that $\mathbf{A}$ is a square matrix of order $n$, we write $\mathbf{A}_{(n)}$. The diagonal elements of $\mathbf{A}_{(n)}$ are $a_{11}, a_{22}, \ldots, a_{nn}$. A square matrix, whose off-diagonal elements are all equal to 0, is called diagonal. The symbols $\mathbf{D}$ and $\Lambda$ always denote diagonal matrices. The identity, or unit, matrix, denoted by $\mathbf{I}$, is a diagonal matrix with all diagonal elements equal to 1. The $i$th column of $\mathbf{I}$ is denoted by

1. Apart from $\Sigma$, all upper-case Greek symbols also denote matrices.

$\mathbf{e}_i$. A vector with all elements equal to 1 is denoted by $\boldsymbol{\iota}$, i.e. $\boldsymbol{\iota}$ is the diagonal of $\mathbf{I}$. A matrix with all elements equal to zero is called a zero, or null, matrix, and is denoted by $\mathbf{0}$. A zero vector is denoted by $\mathbf{o}$. The number of elements of $\mathbf{e}_i$, $\boldsymbol{\iota}$, and $\mathbf{o}$ follows from the context. For instance, we define:

$$\mathbf{E}_{(n)} = \mathbf{I}_{(n)} - \boldsymbol{\iota}(\boldsymbol{\iota}'\boldsymbol{\iota})^{-1}\boldsymbol{\iota}'$$

hence $\boldsymbol{\iota}$ has $n$ elements, so that $(\boldsymbol{\iota}'\boldsymbol{\iota})^{-1} = 1/n$. The $i$th column of a matrix is denoted by the lower-case symbol (with subscript $i$) corresponding to the upper-case symbol which denotes the matrix, with an exception for $\mathbf{I}$, $\mathbf{D}$, $\Lambda$, and $\mathbf{E}$. For instance, $\mathbf{h}_i^*$ is the $i$th column of $\mathbf{H}^*$. But $\mathbf{d}$ and $\boldsymbol{\lambda}$ are the diagonals of $\mathbf{D}$ and $\Lambda$, respectively. Given $\Lambda_{(n)}$ with all diagonal elements $\lambda_1, \lambda_2, \ldots, \lambda_n$ nonnegative, the diagonal matrix with diagonal elements $\lambda_1^{1/2}, \lambda_2^{1/2}, \ldots, \lambda_n^{1/2}$ is denoted by $\Lambda^{1/2}$; if all $\lambda_i$ are strictly positive, then $\lambda_1^{-1/2}, \lambda_2^{-1/2}, \ldots, \lambda_n^{-1/2}$ are the diagonal elements of the matrix denoted by $\Lambda^{-1/2}$.

### *1.A.2. $\mathfrak{M}^n$; length; distance; angle; orthogonal; orthonormal*

The collection of all $n$-element vectors is $\mathfrak{M}^n$, the $n$-dimensional euclidian space. The length of $\mathbf{a}$ is $(\mathbf{a}'\mathbf{a})^{1/2}$. The distance between $\mathbf{a}$ and $\mathbf{b}$ is $[(\mathbf{a}-\mathbf{b})'(\mathbf{a}-\mathbf{b})]^{1/2}$. The cosine of the angle between $\mathbf{a}$ and $\mathbf{b}$, $\mathbf{a} \neq \mathbf{o} \neq \mathbf{b}$, is $(\mathbf{a}'\mathbf{b})(\mathbf{a}'\mathbf{a})^{-1/2}(\mathbf{b}'\mathbf{b})^{-1/2}$; the vectors are orthogonal if $\mathbf{a}'\mathbf{b} = 0$, and orthonormal if $\mathbf{a}'\mathbf{b} = 0$, $\mathbf{a}'\mathbf{a} = \mathbf{b}'\mathbf{b} = 1$.

### *1.A.3. Linear independence; rank; inverse; trace; particular square matrices*

Let $\mathbf{V}$ be an $n \times k$ matrix. If $\mathbf{Va} = \mathbf{o}$ implies $\mathbf{a} = \mathbf{o}$ ($\mathbf{a}$ is a $k$-element vector), then $\mathbf{v}_1, \mathbf{v}_2, \ldots, \mathbf{v}_k$ are linearly independent; otherwise $\mathbf{v}_1, \mathbf{v}_2, \ldots, \mathbf{v}_k$ are linearly dependent. The rank of $\mathbf{V}$, denoted by *rank* $(\mathbf{V})$, is the maximum number of linearly independent columns of $\mathbf{V}$. *Rank* $(\mathbf{V}) =$ *rank* $(\mathbf{V}') =$ *rank* $(\mathbf{V}'\mathbf{V}) =$ *rank* $(\mathbf{VV}') \leq$ *min*$(n, k)$. *Rank* $(\mathbf{AV}) \leq$ *min* $\{$*rank* $(\mathbf{A})$, *rank* $(\mathbf{V})\}$. If $\mathbf{A}_{(n)}$ has rank $n$, $\mathbf{A}$ is called nonsingular (or invertible), and a unique matrix $\mathbf{A}^{-1}$, called the inverse of $\mathbf{A}$, exists such that $\mathbf{A}^{-1}\mathbf{A} = \mathbf{AA}^{-1} = \mathbf{I}$; then *rank* $(\mathbf{AV}) =$ *rank* $(\mathbf{V})$. If *rank* $(\mathbf{A}_{(n)}) =$ *rank* $(\mathbf{V}_{(n)}) = n$, then $(\mathbf{AV})^{-1} = \mathbf{V}^{-1}\mathbf{A}^{-1}$.

The trace of a square matrix $\mathbf{A}$, denoted by $tr(\mathbf{A})$, is the sum of the diagonal elements of $\mathbf{A}$. If $\mathbf{A}$ has order $n \times k$ and $\mathbf{V}$ has order $k \times n$, then $tr(\mathbf{AV}) = tr(\mathbf{VA})$, and $tr(\mathbf{A}'\mathbf{A}) = \sum_{i=1}^{n} \sum_{j=1}^{k} a_{ij}^2$. A square matrix $\mathbf{A}_{(n)}$ is called:

scalar, if $\mathbf{A} = s\mathbf{I}$, where $s$ is a scalar;
symmetric, if $\mathbf{A} = \mathbf{A}'$;
idempotent, if $\mathbf{A} = \mathbf{A}^2$ $(= \mathbf{AA})$;
orthogonal, if $\mathbf{A}' = \mathbf{A}^{-1}$;
lower triangular, if $a_{ij} = 0$ for $i = 1, 2, \ldots, n\text{-}1$ and $j = i+1, i+2, \ldots, n$;
nonnegative definite, if $\mathbf{x}'\mathbf{A}\mathbf{x} \geqslant 0$ for each nonzero vector $\mathbf{x}$;
positive definite, if $\mathbf{x}'\mathbf{A}\mathbf{x} > 0$ for each nonzero vector $\mathbf{x}$.

An $n \times k$ matrix $\mathbf{A}$ satisfying $\mathbf{A}'\mathbf{A} = \mathbf{I}_{(k)}$ is called an orthogonal $n \times k$ matrix. $\mathbf{A}$ is nonnegative definite or positive definite is often abbreviated to $\mathbf{A} \geqslant \mathbf{0}$ or $\mathbf{A} > \mathbf{0}$, respectively.

### *1.A.4. Space; subspace; linear combination; orthogonal complement; basis*

The collection of all $n$-element vectors $\mathbf{b}$, $\mathbf{b} = \mathbf{Va}$, where $\mathbf{V}$ is an $n \times k$ matrix with rank $r$ and $\mathbf{a}$ lies in $\mathcal{M}^k$, is an $r$-dimensional space, denoted by $\mathcal{M}(\mathbf{V})$, which is a subspace of $\mathcal{M}^n$; $\mathbf{b} = a_1\mathbf{v}_1 + a_2\mathbf{v}_2 + \ldots + a_k\mathbf{v}_k$ is called a linear combination of $\mathbf{v}_1, \mathbf{v}_2, \ldots, \mathbf{v}_k$. It is said that $\mathbf{v}_1, \mathbf{v}_2, \ldots, \mathbf{v}_k$ span $\mathcal{M}(\mathbf{V})$. Let the $n \times p$ matrix $\mathbf{A}$ satisfy $\mathbf{V}'\mathbf{A} = \mathbf{0}$. Then every vector in $\mathcal{M}(\mathbf{A})$ is orthogonal to every vector in $\mathcal{M}(\mathbf{V})$, which fact is denoted by $\mathcal{M}(\mathbf{A}) \perp \mathcal{M}(\mathbf{V})$. The orthogonal complement of $\mathcal{M}(\mathbf{V})$, denoted by $\mathcal{M}(\mathbf{V})^{\perp}$, is the collection of all vectors in $\mathcal{M}^n$ that are orthogonal to every vector in $\mathcal{M}(\mathbf{V})$. For instance, if $\mathbf{H}'_{(n)} = \mathbf{H}^{-1}$, then $\mathcal{M}(\mathbf{h}_1) \perp \mathcal{M}(\mathbf{h}_2)$ and $\mathcal{M}(\mathbf{h}_1)^{\perp} = \mathcal{M}(\mathbf{h}_2, \mathbf{h}_3, \ldots, \mathbf{h}_n)$. The orthogonal complement of $\mathcal{M}(\mathbf{V})^{\perp}$ is $\mathcal{M}(\mathbf{V})$.

Assuming that the first $r$ columns of the $n \times k$ matrix $\mathbf{V}$ with rank $r$ are linearly independent, then the last $k - r$ columns of $\mathbf{V}$ are linear combinations of $\mathbf{v}_1, \mathbf{v}_2, \ldots, \mathbf{v}_r$. Hence, $\mathcal{M}(\mathbf{v}_1, \mathbf{v}_2, \ldots, \mathbf{v}_r) = \mathcal{M}(\mathbf{V})$. A basis of $\mathcal{M}(\mathbf{V})$ is a set of linearly independent vectors which spans $\mathcal{M}(\mathbf{V})$. Thus $\mathbf{v}_1, \mathbf{v}_2, \ldots, \mathbf{v}_r$ form a basis of $\mathcal{M}(\mathbf{V})$. If $r = k$ and $\mathbf{A}_{(k)}$ is a nonsingular matrix, then the columns of $\mathbf{VA}$ form a basis of $\mathcal{M}(\mathbf{V})$;

given **VA**, every vector **b** in $\mathcal{M}$(**V**) can be written as **VAg** for some unique vector **g** in $\mathcal{M}^k$.

*1.A.5. Eigenvalue; eigenvector*

Let **A** be a square matrix of order $n$. The scalar $\lambda_i$ is an eigenvalue of **A** and the nonzero vector $\mathbf{h}_i$ is a corresponding eigenvector if and only if $\mathbf{Ah}_i = \lambda_i \mathbf{h}_i$. It follows that $s\mathbf{h}_i$, where $s$ is an arbitrary nonzero scalar, is also an eigenvector corresponding to $\lambda_i$. For convenience, we shall always assume that every real-valued $\mathbf{h}_i$ has a length 1 and that its first nonzero element is positive. **A** has $n$ eigenvalues, which need not all be different; if an eigenvalue occurs $m$ times, it is called multiple with multiplicity $m$; if $m = 1$, it is called distinct. If **A** is symmetric, then:

(1) all eigenvalues are real and the elements of the eigenvectors are real;
(2) if $\lambda_i \neq \lambda_j$, then $\mathbf{h}_i'\mathbf{h}_j = 0$;
(3) $\mathbf{h}_i$ is unique if $\lambda_i$ is distinct;
(4) if $\lambda^*$ has multiplicity $m$, say $\lambda_{i1} = \lambda_{i2} = \ldots = \lambda_{im} = \lambda^*$, then $\mathbf{Ah} = \lambda^*\mathbf{h}$ has $m$ linearly independent solutions – for convenience we take $\mathbf{h}_{i1}, \mathbf{h}_{i2}, \ldots, \mathbf{h}_{im}$ mutually orthogonal.

It follows that every symmetric matrix $\mathbf{A}_{(n)}$ can be written as $\mathbf{A} = \mathbf{H\Lambda H}'$ with $\mathbf{H}' = \mathbf{H}^{-1}$, and we have:

*rank* (**A**) = *rank* ($\Lambda$) = number of nonzero eigenvalues;
*tr* (**A**) = *tr* ($\Lambda$);
if $\mathbf{A} = \mathbf{I}$, then $\Lambda = \mathbf{I}$ and **H** is arbitrary;
if $\mathbf{A} = \mathbf{D}$, then $d_1, d_2, \ldots, d_n$ are the eigenvalues and $\mathbf{H} = \mathbf{I}$ always satisfies;
if *rank* (**A**) = $r$, then $\mathbf{A} = \mathbf{G}\Lambda^*_{(r)}\mathbf{G}'$, where $\lambda_1^*, \lambda_2^*, \ldots, \lambda_r^*$ are the nonzero eigenvalues of **A** and $\mathbf{g}_1, \mathbf{g}_2, \ldots, \mathbf{g}_r$ are corresponding eigenvectors, $\mathbf{G}'\mathbf{G} = \mathbf{I}_{(r)}$ – the set $\mathbf{g}_1, \mathbf{g}_2, \ldots, \mathbf{g}_r$ is an orthonormal basis of $\mathcal{M}$(**A**);
if $\mathbf{A} = \mathbf{A}^2$ and *rank* (**A**) = $r$, then the eigenvalue 1 has multiplicity $r$ and the eigenvalue 0 has multiplicity $n - r$; *rank* (**A**) = *tr* (**A**); $\mathbf{A} = \mathbf{GG}'$ with $\mathbf{G}'\mathbf{G} = \mathbf{I}_{(r)}$;

$\mathbf{A} \geqslant \mathbf{0}$ if and only if $\mathbf{\Lambda} \geqslant \mathbf{0}$; $\mathbf{A}^{1/2}$ is defined as $\mathbf{H\Lambda}^{1/2}\mathbf{H}'$;
$\mathbf{A} > \mathbf{0}$ if and only if $\mathbf{\Lambda} > \mathbf{0}$; $\mathbf{A}^{-1/2} = \mathbf{H\Lambda}^{-1/2}\mathbf{H}'$.

### *1.A.6. Some useful results*

$\mathcal{M}(\mathbf{F}) = \mathcal{M}(\mathbf{FVF}')$ if $\mathbf{V} = \mathbf{V}' > \mathbf{0}$ (or $-\mathbf{V} = -\mathbf{V}' > \mathbf{0}$) (1.A.6.1)

*Proof*: $\mathcal{M}(\mathbf{F})^{\perp}$ is the collection of vectors $\mathbf{a}$ such that $\mathbf{F}'\mathbf{a} = \mathbf{o}$. Then $\mathbf{FVF}'\mathbf{a} = \mathbf{o}$, so that $\mathcal{M}(\mathbf{F})^{\perp}$ lies in $\mathcal{M}(\mathbf{FVF}')^{\perp}$. The latter space is the collection of vectors $\mathbf{b}$ such that $\mathbf{FVF}'\mathbf{b} = \mathbf{o}$. Assuming $\mathbf{V} > \mathbf{0}$, then $\mathbf{0} = \mathbf{b}'\mathbf{FVF}'\mathbf{b} = (\mathbf{V}^{1/2}\mathbf{F}'\mathbf{b})'(\mathbf{V}^{1/2}\mathbf{F}'\mathbf{b})$, which implies $\mathbf{V}^{1/2}\mathbf{F}'\mathbf{b} = \mathbf{o}$. Since $\mathbf{V}^{1/2}$ is nonsingular, we must have $\mathbf{F}'\mathbf{b} = \mathbf{o}$, so that $\mathcal{M}(\mathbf{FVF}')^{\perp}$ lies in $\mathcal{M}(\mathbf{F})^{\perp}$. It follows that $\mathcal{M}(\mathbf{F})^{\perp} = \mathcal{M}(\mathbf{FVF}')^{\perp}$, and hence, $\mathcal{M}(\mathbf{F}) = \mathcal{M}(\mathbf{FVF}')$. Assuming $-\mathbf{V} > \mathbf{0}$, then $\mathbf{FVF}'\mathbf{b} = \mathbf{o}$ if and only if $\mathbf{FWF}'\mathbf{b} = \mathbf{o}$, where $-\mathbf{V} = \mathbf{W} > \mathbf{0}$, and the same result follows.

$\mathcal{M}(\mathbf{W})^{\perp} = \mathcal{M}(\mathbf{I} - \mathbf{W}')$ if $\mathbf{W} = \mathbf{W}^2$ (1.A.6.2)

*Proof*: $\mathcal{M}(\mathbf{W})^{\perp}$ is the collection of vectors $\mathbf{a}$ such that $\mathbf{W}'\mathbf{a} = \mathbf{o}$ or equivalently, $(\mathbf{I} - \mathbf{W}')\mathbf{a} = \mathbf{a}$, so that every vector $\mathbf{a}$ is a linear combination of the columns of $\mathbf{I} - \mathbf{W}'$, i.e. $\mathcal{M}(\mathbf{W})^{\perp}$ lies in $\mathcal{M}(\mathbf{I} - \mathbf{W}')$. Conversely, $\mathcal{M}(\mathbf{I} - \mathbf{W}')$ is the collection of vectors $\mathbf{b}$ such that $\mathbf{b} = (\mathbf{I} - \mathbf{W}')\mathbf{c}$ for some $\mathbf{c}$. Then $\mathbf{W}'\mathbf{b} = \mathbf{W}'(\mathbf{I} - \mathbf{W}')\mathbf{c} = \mathbf{o}$, so that every vector $\mathbf{b}$ lies in $\mathcal{M}(\mathbf{W})^{\perp}$, i.e. $\mathcal{M}(\mathbf{I} - \mathbf{W}')$ lies in $\mathcal{M}(\mathbf{W})^{\perp}$. It follows that $\mathcal{M}(\mathbf{W})^{\perp} = \mathcal{M}(\mathbf{I} - \mathbf{W}')$.

If $\mathbf{FF}' = \mathbf{VV}'$, then $\mathbf{F} = \mathbf{VG}'$ for some $\mathbf{G}$; if $\mathbf{V}'\mathbf{V}$ is nonsingular, then $\mathbf{G}$ is unique and satisfies $\mathbf{G}'\mathbf{G} = \mathbf{I}$. (1.A.6.3)

*Proof*: Using Result 1.A.6.1 twice, we have $\mathcal{M}(\mathbf{V}) = \mathcal{M}(\mathbf{VV}') = \mathcal{M}(\mathbf{FF}') = \mathcal{M}(\mathbf{F})$, so that every column of $\mathbf{F}$ lies in $\mathcal{M}(\mathbf{V})$, i.e. $\mathbf{F} = \mathbf{VG}'$ for some $\mathbf{G}$. $\mathbf{V}'\mathbf{V}$ is nonsingular if and only if the columns of $\mathbf{V}$ are linearly independent. Then $\mathbf{F} = \mathbf{VG}'$ with $\mathbf{G}$ unique, $\mathbf{G}' = (\mathbf{V}'\mathbf{V})^{-1}\mathbf{V}'\mathbf{F}$. Hence, $\mathbf{G}'\mathbf{G} = \mathbf{I}$.

If $\lambda$ is a nonzero eigenvalue of $\mathbf{AV}$ with multiplicity $m$, $m \geqslant 1$, then $\lambda$ is an eigenvalue of $\mathbf{VA}$ with multiplicity $n$, $n \geqslant 1$; if $\mathbf{AV}$ and $\mathbf{VA}$ are symmetric, then $n = m$. (1.A.6.4)

*Proof*: From $\mathbf{AVh} = \lambda\mathbf{h} \neq \mathbf{o}$ it follows that $\mathbf{Vh} \neq \mathbf{o}$ and $\mathbf{VA}(\mathbf{Vh}) =$

$\lambda$ (**Vh**), so that $\lambda$ is an eigenvalue of both **AV** and **VA**. If **AV** is symmetric, then $\mathbf{H} = [\mathbf{h}_1 \vdots \mathbf{h}_2 \vdots \ldots \vdots \mathbf{h}_m]$ in $\mathbf{AVH} = \lambda\mathbf{H}$ satisfies $\mathbf{H}'\mathbf{H} = \mathbf{I}_{(m)}$, and it follows that *rank* (**VH**) = $m$. Hence, $\mathbf{VA}(\mathbf{VH}) = \lambda(\mathbf{VH})$ implies that **VA**, which matrix is symmetric, has eigenvalue $\lambda$ with $n \geqslant m$, since **VH** contains $m$ linearly independent eigenvectors. Similarly, by interchanging **A** and **V**, we obtain $m \geqslant n$, so that we must have $n = m$.

Given $\mathbf{A} = \mathbf{A}'_{(n)} > \mathbf{0}$ and $\mathbf{V} = \mathbf{V}'_{(n)}$, a matrix **F** exists such that $\mathbf{F}'\mathbf{AF} = \mathbf{I}$ and $\mathbf{F}'\mathbf{VF} = \mathbf{D}$, i.e. **F** diagonalizes both **A** and **V**. (1.A.6.5)

*Proof*: The matrix $\mathbf{A}^{-\frac{1}{2}}\mathbf{VA}^{-\frac{1}{2}}$ can be written as $\mathbf{GDG}'$ with $\mathbf{G}' = \mathbf{G}^{-1}$. Defining $\mathbf{F} = \mathbf{A}^{-\frac{1}{2}}\mathbf{G}$, the result follows.

Given $\mathbf{A} = \mathbf{A}'_{(n)} > \mathbf{0}$ and $\mathbf{T}' = \mathbf{T}^{-1}_{(n)}$, then $(\mathbf{TAT}')^{-\frac{1}{2}} = \mathbf{TA}^{-\frac{1}{2}}\mathbf{T}'$ (1.A.6.6)

*Proof*: Let $\mathbf{A} = \mathbf{P\Lambda P}'$ with $\mathbf{P}' = \mathbf{P}^{-1}$. Then $\mathbf{TAT}' = (\mathbf{TP})\,\mathbf{\Lambda}\,(\mathbf{TP})'$ with $(\mathbf{TP})' = (\mathbf{TP})^{-1}$, so that $(\mathbf{TAT}')^{-\frac{1}{2}} = (\mathbf{TP})\mathbf{\Lambda}^{-\frac{1}{2}}(\mathbf{TP})' = \mathbf{T}(\mathbf{P\Lambda}^{-\frac{1}{2}}\mathbf{P}')\mathbf{T}' = \mathbf{TA}^{-\frac{1}{2}}\mathbf{T}'$.

Let **A** be an $m \times r$ matrix with rank $s$, $s \leqslant r \leqslant m$, and $\mathbf{A}'\mathbf{A} = \mathbf{T\Lambda T}'$ with $\mathbf{T}' = \mathbf{T}^{-1}$. Then $\mathbf{A} = \mathbf{U\Lambda}^{\frac{1}{2}}\mathbf{T}'$ for some **U** satisfying $\mathbf{U}'\mathbf{U} = \mathbf{I}_{(r)}$; given **T** and $\mathbf{\Lambda}$, $s$ columns of **U** are unique. (1.A.6.7)

*Proof*: Let $\lambda_{s+1} = \ldots = \lambda_r = 0$, then:

$$\mathbf{A}'\mathbf{A} = \mathbf{T\Lambda T}' = [\mathbf{T}_1 \vdots \mathbf{T}_2]\begin{bmatrix}\mathbf{\Lambda}^*_{(s)} & \vdots & \mathbf{0} \\ \cdots & \cdots & \cdots \\ \mathbf{0} & \vdots & \mathbf{0}_{(r\text{-}s)}\end{bmatrix}\begin{bmatrix}\mathbf{T}'_1 \\ \cdots \\ \mathbf{T}'_2\end{bmatrix} = \mathbf{T}_1\mathbf{\Lambda}^*_{(s)}\mathbf{T}'_1$$

$$= (\mathbf{T}_1\mathbf{\Lambda}^{*\frac{1}{2}}_{(s)})(\mathbf{T}_1\mathbf{\Lambda}^{*\frac{1}{2}}_{(s)})'$$

where $\lambda^*_i = \lambda_i > 0$ for $i = 1, 2, \ldots, s$. In accordance with Result 1.A.6.3, $\mathbf{A}' = \mathbf{T}_1\mathbf{\Lambda}^{*\frac{1}{2}}_{(s)}\mathbf{U}'_1$ for some unique $m \times s$ matrix $\mathbf{U}_1$ satisfying $\mathbf{U}'_1\mathbf{U}_1 = \mathbf{I}_{(s)}$. Let $\mathbf{U}_2$ be an $m \times (r\text{-}s)$ matrix consisting of $r$-$s$ orthonormal column vectors lying in the ($m$-$s$)-dimensional space $\mathcal{M}(\mathbf{U}_1)^{\perp}$, then:

$$\mathbf{A} = \mathbf{U}_1\mathbf{\Lambda}^{*\frac{1}{2}}_{(s)}\mathbf{T}'_1 = [\mathbf{U}_1 \vdots \mathbf{U}_2]\begin{bmatrix}\mathbf{\Lambda}^{*\frac{1}{2}}_{(s)} & \vdots & \mathbf{0} \\ \cdots & \cdots & \cdots \\ \mathbf{0} & \vdots & \mathbf{0}_{(r\text{-}s)}\end{bmatrix}\begin{bmatrix}\mathbf{T}'_1 \\ \cdots \\ \mathbf{T}'_2\end{bmatrix} = \mathbf{U\Lambda}^{\frac{1}{2}}\mathbf{T}'$$

where $\mathbf{U} = [\mathbf{U}_1 \vdots \mathbf{U}_2]$, $\mathbf{U}'\mathbf{U} = \mathbf{I}_{(r)}$.

Let $\mathbf{A} = \mathbf{U}\Lambda^{1/2}\mathbf{T}'$ as in Result 1.A.6.7. Then, for all $m \times r$ matrices $\mathbf{H}$ satisfying $\mathbf{H}'\mathbf{H} = \mathbf{I}_{(r)}$, we have:

$$tr(\mathbf{H}'\mathbf{A}) \leqslant tr(\mathbf{H}_0'\mathbf{A}) = tr(\Lambda^{1/2})$$

where $\mathbf{H}_0 = \mathbf{U}\mathbf{T}'$; if $s = r$, then $\mathbf{H}_0 = \mathbf{A}(\mathbf{A}'\mathbf{A})^{-1/2}$ and:

$$tr(\mathbf{H}'\mathbf{A}) < tr(\mathbf{H}_0'\mathbf{A})$$

for every $\mathbf{H} \neq \mathbf{H}_0$. (1.A.6.8)

*Proof*: Verify that $\mathbf{H}_0'\mathbf{H}_0 = \mathbf{T}\mathbf{U}'\mathbf{U}\mathbf{T}' = \mathbf{I}$, that $tr(\mathbf{H}_0'\mathbf{A}) = tr(\mathbf{T}\mathbf{U}'\mathbf{U}\Lambda^{1/2}\mathbf{T}') = tr(\Lambda^{1/2}\mathbf{T}'\mathbf{T}) = tr(\Lambda^{1/2})$, and that $\mathbf{A}(\mathbf{A}'\mathbf{A})^{-1/2} = \mathbf{U}\Lambda^{1/2}\mathbf{T}'\mathbf{T}\Lambda^{-1/2}\mathbf{T}' = \mathbf{U}\mathbf{T}'$ if $\Lambda^{-1}$ exists. Any matrix $\mathbf{H}$ can be written as $\mathbf{H} = \mathbf{H}_0 + \mathbf{F}$, with $\mathbf{F}$ satisfying:

$$\mathbf{0} = \mathbf{H}'\mathbf{H} - \mathbf{H}_0'\mathbf{H}_0 = \mathbf{F}'\mathbf{U}\mathbf{T}' + \mathbf{T}\mathbf{U}'\mathbf{F} + \mathbf{F}'\mathbf{F}$$

We wish to prove that $tr(\mathbf{H}'\mathbf{A}) - tr(\mathbf{H}_0'\mathbf{A}) = tr(\mathbf{F}'\mathbf{A}) \leqslant 0$. From the condition on $\mathbf{F}$ we find:

$$tr(\mathbf{T}\Lambda^{1/2}\mathbf{T}'\mathbf{F}'\mathbf{U}\mathbf{T}') + tr(\mathbf{T}\Lambda^{1/2}\mathbf{T}'\mathbf{T}\mathbf{U}'\mathbf{F}) + tr(\mathbf{T}\Lambda^{1/2}\mathbf{T}'\mathbf{F}'\mathbf{F}) = 0$$

Applying $tr(\mathbf{V}\mathbf{Y}) = tr(\mathbf{Y}\mathbf{V}) = tr(\mathbf{V}'\mathbf{Y}')$, we find

$$tr(\mathbf{F}'\mathbf{A}) = -\tfrac{1}{2}tr(\mathbf{F}\mathbf{T}\Lambda^{1/2}\mathbf{T}'\mathbf{F}') \leqslant 0$$

The inequality follows from $\Lambda^{1/2} \geqslant \mathbf{0}$. Clearly, $tr(\mathbf{F}'\mathbf{A}) = 0$ if and only if $\mathbf{F}\mathbf{T}\Lambda = \mathbf{0}$. If $s = r$, then $\mathbf{F}\mathbf{T}\Lambda = \mathbf{0}$ implies $\mathbf{F} = \mathbf{0}$, since $\mathbf{T}\Lambda$ is non-singular. If $s < r$, then at least one of the $\lambda_i$ is zero, say $\lambda_r = 0$, so that the last column of $\mathbf{T}\Lambda$ is a zero column. The matrix $\mathbf{F}$, whose last row is equal to the last column of $\mathbf{T}$ and with all other elements equal to zero, satisfies $\mathbf{F}\mathbf{T}\Lambda = \mathbf{0}$, so that $tr(\mathbf{H}'\mathbf{A}) = tr(\mathbf{H}_0'\mathbf{A})$ does not imply $\mathbf{H} = \mathbf{H}_0$ when $s < r$.

# 2. Tabulable quadratic ratio tests

## 2.1. The form of the test statistic $T$

We deal with tests for hypotheses concerning parameters of $\Gamma$. For instance, it is assumed that the $(i, j)$th element of $\Gamma$ is equal to $\rho^{|i-j|}$ and the null hypothesis $\mathcal{H}_0$ is: $\rho = 0$, while the alternative hypothesis $\mathcal{H}_A$ is: $\rho = \rho^* > 0$. On the basis of an observed sample, a test is carried out to decide whether to accept $\mathcal{H}_0$ or to reject $\mathcal{H}_0$ in favour of $\mathcal{H}_A$.

In accordance with (1.1), the vector y has a normal distribution with mean $\mathbf{X\beta}$ and covariance matrix $\sigma^2\Gamma$. The sample space of y is divided into two mutually exclusive subspaces, the critical or rejection region and the acceptance region. If a sample, i.e. an observation vector y, falls within the critical region, then $\mathcal{H}_0$ must be rejected; otherwise, $\mathcal{H}_0$ must be accepted. Clearly, the critical region defines the test.

Since we are dealing with $n$-dimensional normal sample distributions, both under $\mathcal{H}_0$ and $\mathcal{H}_A$, each nondegenerate critical region involves positive probabilities of taking a wrong decision: $P(I) = \alpha$, the probability of rejecting $\mathcal{H}_0$ when $\mathcal{H}_0$ is true, also called the significance level of the test; and $P(II)$, the probability of accepting $\mathcal{H}_0$ when $\mathcal{H}_A$ is true. $1 - P(II)$ is the power of the test.

Given $\mathcal{H}_0$ and $\mathcal{H}_A$, several critical regions may be proposed, each region defining a test. The *best* critical region is the one associated with maximal power for fixed significance level $\alpha$; the corresponding test is called *most powerful* (*MP*). Sometimes the best critical region is the same for a certain set of alternative hypotheses. In that case the corresponding test is called a *uniformly most powerful* (*UMP*) test with respect to that set of alternatives.

A *UMP* test is very attractive: the performance of just one testing procedure is sufficient to test against each of the alternative hypotheses included in the set of alternatives, and this single testing procedure involves minimal probabilities of taking a wrong decision in all cases. For instance, if there is a most powerful test for $\mathcal{H}_0 : \rho = 0$ against $\mathcal{H}_A : \rho = \rho^* > 0$, and the critical region does not depend on $\rho^*$, then this test is a *UMP* test for $\mathcal{H}_0 : \rho = 0$ against $\mathcal{H}_A : \rho > 0$ (the situation would be awful if $\mathcal{H}_A : \rho = 0.6$ and $\mathcal{H}_A : \rho = 0.7$ and $\mathcal{H}_A :$ $\rho = 0.8$ require three different tests). Here one cannot speak of *the* power of the test. Instead we have a *power function* of $\rho$. This is the probability that the observation vector **y** falls within the critical region, which probability depends on $\rho$. In particular, for $\rho = 0$ this probability is the significance level.

When the probability distribution of a random sample is specified completely, both under $\mathcal{H}_0$ and $\mathcal{H}_A$, then the well-known Neyman-Pearson lemma provides a systematic method of determining a best critical region. However, the distribution of **y** depends on the unknown $\boldsymbol{\beta}$ and $\sigma^2$, so we cannot use the Neyman-Pearson lemma.

A region in the sample space of **y** is said to be a *similar region* with respect to $\boldsymbol{\beta}$ and $\sigma^2$, when the probability that a sample falls within that region is independent of $\boldsymbol{\beta}$ and $\sigma^2$. Consider a vector $\mathbf{w} = \mathbf{B}'\mathbf{y}$, where **B** is a matrix independent of **y** and satisfying $\mathbf{B}'\mathbf{X} = \mathbf{0}$.

Any region which is defined exclusively in terms of $\mathbf{w}/\sqrt{\mathbf{w}'\mathbf{C}\mathbf{w}}$, where **C** is some nonnegative definite matrix, is a similar region with respect to $\boldsymbol{\beta}$ and $\sigma^2$. This follows from the fact that $\frac{1}{\sigma}\mathbf{w} = \frac{1}{\sigma}\mathbf{B}'\mathbf{y} \sim \mathcal{N}(\mathbf{o}, \mathbf{B}'\boldsymbol{\Gamma}\mathbf{B})$, so that the distribution of $(\frac{1}{\sigma}\mathbf{w})/\sqrt{(\frac{1}{\sigma}\mathbf{w})'\mathbf{C}(\frac{1}{\sigma}\mathbf{w})} = \mathbf{w}/\sqrt{\mathbf{w}'\mathbf{C}\mathbf{w}}$ cannot depend on $\boldsymbol{\beta}$ and $\sigma^2$. In this study we consider test statistics of the form:

$$T = \frac{\mathbf{w}'\mathbf{A}\mathbf{w}}{\mathbf{w}'\mathbf{C}\mathbf{w}} \; [= \left(\frac{\mathbf{w}}{\sqrt{\mathbf{w}'\mathbf{C}\mathbf{w}}}\right)' \mathbf{A} \left(\frac{\mathbf{w}}{\sqrt{\mathbf{w}'\mathbf{C}\mathbf{w}}}\right)] \qquad (2.1)$$

Under special conditions, a test using a test statistic of the form $T$ is *UMPS* (i.e. *UMP* within the class of *similar* tests). This has been shown by Anderson (1948), who considered the following framework. Suppose that $\Gamma^{-1}$ can be written as:

$$\Gamma^{-1} = c_0 \Psi + \tau \Theta$$

where $c_0 > 0$, $\Psi$ and $\Theta$ are fixed symmetric matrices, $\Psi$ positive definite. The test at significance level $\alpha$ for $\mathcal{H}_0 : \tau = 0$ against $\mathcal{H}_A :$ $\tau > 0$ with the critical region defined by:

$$T^* = \frac{\mathbf{u}^{*\prime}\Theta\mathbf{u}^*}{\mathbf{u}^{*\prime}\Psi\mathbf{u}^*} \leqslant t \qquad (2.2)$$

with $t$ determined by $Pr[T^* \leqslant t \mid \mathcal{H}_0] = \alpha$ and $\mathbf{u}^* = \mathbf{M}^*\mathbf{y} = [\mathbf{I} - \mathbf{X}(\mathbf{X}'\Psi\mathbf{X})^{-1}\mathbf{X}'\Psi]\mathbf{y}$, is *UMPS* if the $k$ columns of $\mathbf{X}$ are linear combinations of $k$ columns of $\mathbf{S}$, where $\mathbf{S}$ is an $n \times n$ matrix such that $\mathbf{S}'\Psi\mathbf{S} = \mathbf{I}_{(n)}$ and $\mathbf{S}'\Theta\mathbf{S} = \mathbf{D}$ (by convention, $\mathbf{D}$ denotes a diagonal matrix, see Section 1.A.1).

Note that the restriction on $\mathbf{X}$ means that $\mathbf{X}$ can be written as $\mathbf{SJG}$, where $\mathbf{G}$ is a nonsingular $k \times k$ matrix and $\mathbf{J}$ is an $n \times k$ matrix consisting of $k$ columns from $\mathbf{I}_{(n)}$, and that $\mathbf{s}_j$ (by convention, the $j$th column of $\mathbf{S}$) $= \Psi^{1/2}\bar{\mathbf{s}}_j$, where $\bar{\mathbf{s}}_j$, $j = 1, 2, \ldots, n$, are orthonormal eigenvectors of $\Psi^{-1/2}\Theta\Psi^{-1/2}$, see Result 1.A.6.5. In practical cases with $\Psi = \mathbf{I}$, the restriction means that the $k$ columns of $\mathbf{X}$ are linear combinations of $k$ eigenvectors of $\Theta$. Anderson also considered the more general decomposition:

$$\Gamma^{-1} = c_0\Psi + \sum_{i=1}^{p} c_i\Theta_i$$

where $\Theta_1, \Theta_2, \ldots, \Theta_p$ are fixed symmetric matrices. The test at significance level $\alpha$ for $\mathcal{H}_0 : \mathbf{c} = \mathbf{o}$ ($\mathbf{c}$ denotes the vector $[c_1\ c_2 \ldots c_p]'$) against $\mathcal{H}_A : \mathbf{c} = \mathbf{c}^*$ ($\mathbf{c}^*$ is a fixed $p$-element vector) with the critical region defined by:

$$T_1^* = \frac{\sum_{i=1}^{p} c_i^*\mathbf{u}^{*\prime}\Theta_i\mathbf{u}^*}{\mathbf{u}^{*\prime}\Psi\mathbf{u}^*} \leqslant t_1$$

with $t_1$ determined by $Pr[T_1^* \leqslant t_1 \mid \mathcal{H}_0] = \alpha$, is *MPS* if the $k$ columns of $\mathbf{X}$ are linear combinations of $k$ vectors characteristic to $\Theta_1, \Theta_2, \ldots, \Theta_p$ in the metric $\Psi$. The last expression suggests that some relation between $\Theta_1, \Theta_2, \ldots, \Theta_p$ should exist. However, defining $\Theta = \sum_{i=1}^{p} c_i^*\Theta_i$, then $\Theta$ is a fixed matrix, and, in accordance with the test defined by (2.2), the condition that the $k$ columns of $\mathbf{X}$ are linear combinations of $k$ columns of $\mathbf{S}$, $\mathbf{S}$ satisfying $\mathbf{S}'\Psi\mathbf{S} = \mathbf{I}_{(n)}$ and $\mathbf{S}'\Theta\mathbf{S} = \mathbf{S}'(\sum_{i=1}^{p} c_i^*\Theta_i)\mathbf{S} = \mathbf{D}$, is sufficient for the test in terms of $T_1^*$ to be *UMPS* against $\mathcal{H}_A : \mathbf{c} = \tau\mathbf{c}^*$ with $\tau > 0$ and $\mathbf{c}^*$ fixed, i.e. the test is *UMPS* if the set of alternatives lie on a straight line in the $\mathbf{c}$-space at one side of the origin. If we would consider another alternative, say $\mathcal{H}_A : \mathbf{c} = \mathbf{c}^{**}$, such that $\mathbf{c}^{**} \neq \tau\mathbf{c}^*$ for all $\tau > 0$, then the *MPS* test against $\mathcal{H}_A : \mathbf{c} = \mathbf{c}^{**}$ reads in terms of:

$$T_2^* = \frac{\sum_{i=1}^{p} c_i^{**} \mathbf{u}^{*\prime} \Theta_i \mathbf{u}^*}{\mathbf{u}^{*\prime} \Psi \mathbf{u}^*} \leqslant t_2$$

which defines another critical region than $T_1^* \leqslant t_1$. When we have two critical regions, each defining a *MPS* test (one against $\mathbf{c} = \mathbf{c}^*$ and one against $\mathbf{c} = \mathbf{c}^{**}$), and the critical regions are not identical, then no *UMPS* test against both alternatives exists. Anderson provided the following illustration, which is of particular interest for us. Let $\Gamma$ be given in (1.5). Then, with $c_0 = 1/(1 - \rho^2)$:

$$\frac{1}{c_0}\Gamma^{-1} = \begin{bmatrix} 1 & -\rho & 0 & 0 & \cdots & 0 & 0 \\ -\rho & 1+\rho^2 & -\rho & 0 & \cdots & 0 & 0 \\ 0 & -\rho & 1+\rho^2 & -\rho & \cdots & 0 & 0 \\ \vdots & & & & & & \vdots \\ 0 & 0 & 0 & 0 & \cdots & 1+\rho^2 & -\rho \\ 0 & 0 & 0 & 0 & \cdots & -\rho & 1 \end{bmatrix}$$

$$= \mathbf{I}_{(n)} - \rho \begin{bmatrix} 0 & 1 & 0 & 0 & \cdots & 0 & 0 \\ 1 & 0 & 1 & 0 & \cdots & 0 & 0 \\ 0 & 1 & 0 & 1 & \cdots & 0 & 0 \\ \vdots & & & & & & \vdots \\ 0 & 0 & 0 & 0 & \cdots & 0 & 1 \\ 0 & 0 & 0 & 0 & \cdots & 1 & 0 \end{bmatrix} + \rho^2 \begin{bmatrix} 0 & 0 & 0 & 0 & \cdots & 0 & 0 \\ 0 & 1 & 0 & 0 & \cdots & 0 & 0 \\ 0 & 0 & 1 & 0 & \cdots & 0 & 0 \\ \vdots & & & & & & \vdots \\ 0 & 0 & 0 & 0 & \cdots & 1 & 0 \\ 0 & 0 & 0 & 0 & \cdots & 0 & 0 \end{bmatrix}$$

$$\equiv \Psi + \frac{c_1}{c_0}\Theta_1 + \frac{c_2}{c_0}\Theta_2 \qquad (2.3)$$

Since the alternatives $[c_1 c_2] = c_0[-\rho \; \rho^2]$ do not lie on a straight line if we consider $\rho > 0$, it follows that no *UMPS* test for $\mathcal{H}_0 : \rho = 0$ against $\mathcal{H}_A : \rho > 0$ exists (see Anderson 1971, sections 6.3. and 6.6). Durbin and Watson (1950, 1951) considered a $\Gamma^{-1}$-matrix which slightly differs from (2.3), namely:

$$\frac{1}{c_0}\Gamma^{-1} = (1 - \rho)^2\mathbf{I} + \rho\mathbf{A}_d$$

where:

$$
\mathbf{A}_d = \begin{bmatrix} 1 & -1 & 0 & 0 & \cdots & 0 & 0 \\ -1 & 2 & -1 & 0 & \cdots & 0 & 0 \\ 0 & -1 & 2 & -1 & \cdots & 0 & 0 \\ \vdots & & & & & & \vdots \\ 0 & 0 & 0 & 0 & \cdots & 2 & -1 \\ 0 & 0 & 0 & 0 & \cdots & -1 & 1 \end{bmatrix} \qquad (2.4)
$$

In virtue of Anderson's results, the test for $\mathcal{H}_0 : \rho = 0$ against $\mathcal{H}_A : \rho > 0$ with the critical region defined by:

$$
d = \frac{\hat{\mathbf{u}}'\mathbf{A}_d\hat{\mathbf{u}}}{\hat{\mathbf{u}}'\hat{\mathbf{u}}} \leqslant d^* \qquad (2.5)
$$

with $d^*$ determined by $Pr[d \leqslant d^* \mid \mathcal{H}_0] = \alpha$, is *UMPS* at significance level $\alpha$ if $\mathbf{X}$ consists of linear combinations of $k$ eigenvectors of $\mathbf{A}_d$. Note that $\mathbf{u}^* = \hat{\mathbf{u}}$, since $\Gamma = \mathbf{I}$ under $\mathcal{H}_0$, and that the coefficient $(1\text{-}\rho)^2$ in $\frac{1}{c_0}\Gamma^{-1} = (1\text{-}\rho)^2\mathbf{I} + \rho\mathbf{A}_d$ is irrelevant in view of $\frac{(1\text{-}\rho^2)}{(1\text{-}\rho)^2}\Gamma^{-1} = \mathbf{I} + \frac{\rho}{(1\text{-}\rho)^2}\mathbf{A}_d$, while $\frac{\rho}{(1\text{-}\rho)^2} > 0$ if and only if $\rho > 0$. This test, with $\mathcal{H}_A : \rho > 0$ in this structure of $\Gamma$ replaced by $\mathcal{H}_A : \rho > 0$ in (2.3), is the *exact Durbin-Watson test*.

In Durbin and Watson (1971) a theoretical reconsideration of the power of the test based on $d$ is presented, namely through the theory of invariance. Considering $\Gamma^{-1} = (1\text{-}\rho)^2\,\mathbf{I} + \rho\mathbf{A}_d$ with $\mathcal{H}_0 : \rho = 0$ and $\mathcal{H}_A : \rho = \rho_1 > 0$, the most powerful invariant test has a critical region of the form:

$$
\bar{d} = \frac{\tilde{\mathbf{u}}\,\mathbf{A}_d\tilde{\mathbf{u}}}{\tilde{\mathbf{u}}\,\tilde{\mathbf{u}}} \leqslant \bar{d}^*
$$

where $\tilde{\mathbf{u}}$ is the *g.l.s.* residual vector under $\mathcal{H}_A : \rho = \rho_1$, i.e. $\tilde{\mathbf{u}} = [\mathbf{I} - \mathbf{X}(\mathbf{X}\ \Gamma_A^{-1}\mathbf{X})^{-1}\mathbf{X}'\Gamma_A^{-1}]\mathbf{y}$ and $\Gamma_A^{-1} = (1\text{-}\rho_1)^2\mathbf{I} + \rho_1\mathbf{A}_d$. For general $\mathbf{X}$ the region depends on $\rho_1$, so that a *UMPI (I = invariant)* test against a suitable family of alternatives does not exist in general. However, when the columns of $\mathbf{X}$ are linear combinations of $k$ eigenvectors of $\mathbf{A}_d$, one of these being the constant term vector, then $\tilde{\mathbf{u}} = \hat{\mathbf{u}}$, so that $\bar{d} = d$, and the critical region does not depend on $\rho_1$; hence, the *UMPS* test against $\mathcal{H}_A : \rho > 0$ is also *UMPI*. For general $\mathbf{X}$ the test based on $d$ is locally most powerful invariant in the neighbourhood of $\mathcal{H}_0$, since, if $\rho_1 \to 0$, then $\tilde{\mathbf{u}} \to \hat{\mathbf{u}}$ and $\bar{d} \to d$.

Berenblut and Webb (1973) proposed the test statistic $g$, see (2.6), yielding tests which are locally most powerful invariant in the neighbourhood of $\mathcal{H}_A : \rho = 1$ for general **X** and *UMPI* for a particular class of **X**-matrices. They compare power functions of the tests for $\mathcal{H}_0 : \rho = 0$ against $\mathcal{H}_A : \rho > 0$ based on $d$ and on $g$, in both the nonstationary model and the stationary model (1.4). $\Gamma^{-1}$ in the stationary model is given in (2.3), and $\Gamma^{-1}$ in the nonstationary model differs from (2.3) in the element (1,1) only, which is $1+\rho^2$ instead of 1. Denoting the latter $\Gamma^{-1}$-matrix with $\rho = 1$ by $\mathbf{A}_b$, then:

$$g = \frac{\tilde{\mathbf{u}}'\mathbf{A}_b\tilde{\mathbf{u}}}{\hat{\mathbf{u}}'\hat{\mathbf{u}}} \tag{2.6}$$

where $\tilde{\mathbf{u}} = [\mathbf{I} - \mathbf{X}(\mathbf{X}'\mathbf{A}_b\mathbf{X})^{-1}\mathbf{X}'\mathbf{A}_b]\mathbf{y}$. In five of the six applications, the differences between the power functions of $d$ and $g$ are negligible for $0 < \rho < 0.7$, and (roughly interpreting the figures) the powers for $g$ are 1.05 times the powers for $d$ at $\rho = 0.9$.

This brief review of most powerful similar tests and (locally) most powerful invariant tests, suggests that (in a power sense) a good test for $\mathcal{H}_0 : \Gamma = \Gamma_0$ against $\mathcal{H}_A : \Gamma = \Gamma_A$ should be based on a test statistic which is a ratio of quadratic forms in *BLU* disturbance estimators, like $T$ in (2.1) with $\mathbf{w}$ equal to $\hat{\mathbf{u}}$, $\mathbf{u}^*$ or $\tilde{\mathbf{u}}$.

## 2.2. Calculable distribution functions

We now examine the probability distribution function $\mathcal{F}(t)$ of $T$:

$$T = \mathbf{w}'\mathbf{A}\mathbf{w}/\mathbf{w}'\mathbf{C}\mathbf{w} \tag{2.7}$$

where **A** and **C** are fixed symmetric $p \times p$ matrices, **C** nonnegative definite, and the $p$-element vector **w** is a linear function of **y**:

$$\mathbf{w} = \mathbf{B}'\mathbf{y} \tag{2.8}$$

where **B** has order $n \times p$ and rank $r$. We assume that $r$ is large, preferably as large as possible, because of the following argument. $T$ is to be used as a test statistic and all information, on the basis of which we perform a test, is contained in the sample **y**. This information in **w** generally decreases as $r$ becomes smaller. One may expect therefore, that the larger $r$, the more powerful the test.

$T$ is not defined if its denominator is zero, i.e. $\mathbf{y}'\mathbf{BCB}'\mathbf{y} = 0$. Given the high rank of **B**, we have $\mathbf{BCB}' = \mathbf{0}$ only if the rank of **C** is very small. This does not happen in our applications. Usually $\mathbf{BCB}'$ is singular. For instance, if $\mathbf{C} = \mathbf{I}_{(n)}$ and $\mathbf{w} = \hat{\mathbf{u}}$, so that $\mathbf{B}' = \mathbf{M}$, then $\mathbf{BCB}' = \mathbf{M}$ and the denominator of $T$ is $\hat{\mathbf{u}}'\hat{\mathbf{u}} = \mathbf{y}'\mathbf{My}$, $rank\,(\mathbf{M}) = n\text{-}k < n$. Hence, vectors **y** exist such that $\mathbf{y}'\mathbf{BCB}'\mathbf{y} = 0$; in particular, $\mathbf{y}'\mathbf{My} = 0$ if and only if **y** is a linear combination of the columns of **X**, implying $\hat{\mathbf{u}} = \mathbf{o}$. It can be shown that $Pr[\mathbf{y}'\mathbf{BCB}'\mathbf{y} > 0] = 1$. $\mathcal{F}(t)$ can be written as:

$$\mathcal{F}(t) = Pr[T \leqslant t] = Pr[\mathbf{w}'\mathbf{Aw} \leqslant t\mathbf{w}'\mathbf{Cw}] = Pr[\mathbf{w}'(\mathbf{A} - t\mathbf{C})\mathbf{w} \leqslant 0] \tag{2.9}$$

We first express $\mathcal{F}(t)$ as the probability that a weighted sum of mutually independent $\chi^2(1)$-variables is negative. The $p \times p$ matrix $\mathbf{B}'\mathbf{\Gamma B}$ is symmetric nonnegative definite with rank $r$ (see Result 1.A.6.1), so that $\mathbf{B}'\mathbf{\Gamma B} = \mathbf{FDF}'$, where the $p \times r$ matrix **F** satisfies $\mathbf{F}'\mathbf{F} = \mathbf{I}_{(r)}$ and the $r \times r$ matrix **D** (by convention, **D** and $\mathbf{\Lambda}$ denote diagonal matrices, see Section 1.A.1) is positive definite. Using Results 1.A.6.1 and 1.A.6.3, we have $\mathcal{M}(\mathbf{B}') = \mathcal{M}(\mathbf{FD}^{1/2})$. From $\mathbf{w} = \mathbf{B}'\mathbf{y}$ it follows that **w** lies in $\mathcal{M}(\mathbf{B}')$, so that $\mathbf{w} = \mathbf{FD}^{1/2}\mathbf{v}$ for some unique $r$-element vector **v**. Given $\mathbf{y} \sim \mathcal{N}(\mathbf{X}\boldsymbol{\beta}, \sigma^2\mathbf{\Gamma})$, we have $\mathbf{v} = \mathbf{D}^{-1/2}\mathbf{F}'\mathbf{w} = \mathbf{D}^{-1/2}\mathbf{F}'\mathbf{B}'\mathbf{y} \sim \mathcal{N}(\mathbf{D}^{-1/2}\mathbf{F}'\mathbf{B}'\mathbf{X}\boldsymbol{\beta}, \sigma^2\mathbf{I}_{(r)})$. Let $\mathbf{D}^{1/2}\mathbf{F}'(\mathbf{A} - t\mathbf{C})\mathbf{FD}^{1/2} = \mathbf{L\Lambda L}'$ with $\mathbf{L}' = \mathbf{L}^{-1}$ and $\lambda_1 \leqslant \lambda_2 \leqslant \ldots \leqslant \lambda_r$, and define $\mathbf{z} = \frac{1}{\sigma}\mathbf{L}'\mathbf{v}$, so that $\mathbf{z} \sim \mathcal{N}(\frac{1}{\sigma}\mathbf{L}'\mathbf{D}^{-1/2}\mathbf{F}'\mathbf{B}'\mathbf{X}\boldsymbol{\beta}, \mathbf{I}_{(r)})$, then:

$$\mathbf{w}'(\mathbf{A} - t\mathbf{C})\mathbf{w} = \mathbf{v}'\mathbf{D}^{1/2}\mathbf{F}'(\mathbf{A} - t\mathbf{C})\mathbf{FD}^{1/2}\mathbf{v} = \mathbf{v}'\mathbf{L\Lambda L}'\mathbf{v} = \sigma^2\mathbf{z}'\mathbf{\Lambda z}$$

Hence, in accordance with (2.9):

$$\mathcal{F}(t) = Pr[\mathbf{w}'(\mathbf{A} - t\mathbf{C})\mathbf{w} \leqslant 0] = Pr[\mathbf{z}'\mathbf{\Lambda z} \leqslant 0] = Pr[\sum_{i=1}^{r} \lambda_i z_i^2 \leqslant 0] \tag{2.10}$$

where $\lambda_1 \leqslant \lambda_2 \leqslant \ldots \leqslant \lambda_r$ are the eigenvalues of $\mathbf{D}^{1/2}\mathbf{F}'(\mathbf{A} - t\mathbf{C})\mathbf{FD}^{1/2}$, and $z_1^2, z_2^2, \ldots, z_r^2$ are mutually independent noncentral $\chi^2(1)$-variables with noncentrality parameters $\delta_1^2, \delta_2^2, \ldots, \delta_r^2$; **F** and **D** are matrices such that $\mathbf{F}'\mathbf{F} = \mathbf{I}_{(r)}$ and $\mathbf{FDF}' = \mathbf{B}'\mathbf{\Gamma B}$; $\boldsymbol{\delta} = [\delta_1\ \delta_2 \ldots \delta_r]' = \frac{1}{\sigma}\mathbf{L}'\mathbf{D}^{-1/2}\mathbf{F}'\mathbf{B}'\mathbf{X}\boldsymbol{\beta}$, where **L** is a square orthogonal matrix such that $\mathbf{L\Lambda L}' = \mathbf{D}^{1/2}\mathbf{F}'(\mathbf{A} - t\mathbf{C})\mathbf{FD}^{1/2}$.

At present two methods are available to calculate a probability like (2.10), provided that $\lambda_1, \lambda_2, \ldots, \lambda_r$ are individually known (apart from simultaneous multiplication by a positive number), and that each of the numbers of degrees of freedom and the noncentrality parameters of $z_1^2, z_2^2, \ldots, z_r^2$ is known. These methods are the Imhof (1961) procedure as adapted in Koerts and Abrahamse (1969),[1] and the procedure of Pan Jie-jian (1968). Let $z_i^2\{\chi^2(f_i, \theta_i)\}$denote that $z_i^2$ has a $\chi^2$-distribution with $f_i$ degrees of freedom and noncentrality parameter $\theta_i$; then $Pr[\sum_{i=1}^{r} \lambda_i z_i^2 \{\chi^2(1,0)\} \leqslant 0]$ can be calculated by means of the procedure of Pan Jie-jian, given $\lambda_1, \lambda_2, \ldots, \lambda_r$; whereas $Pr[\sum_{i=1}^{r} \lambda_i z_i^2 \{\chi^2(f_i, \theta_i)\} \leqslant c]$ can be calculated by means of the procedure of Imhof, given $\lambda_1, f_1, \theta_1, \lambda_2, f_2, \theta_2, \ldots, \lambda_r, f_r, \theta_r$, and $c$; in both cases the $\chi^2$-variables are assumed to be mutually stochastically independent.

The noncentrality parameter $\delta_i^2$ of $z_i^2$ in (2.10) is equal to:

$$\delta_i^2 = \frac{1}{\sigma^2} \boldsymbol{\beta}'\mathbf{X}'\mathbf{B}\mathbf{F}\mathbf{D}^{-\frac{1}{2}}\mathbf{l}_i\mathbf{l}_i'\mathbf{D}^{-\frac{1}{2}}\mathbf{F}'\mathbf{B}'\mathbf{X}\boldsymbol{\beta} \geqslant 0$$

where $\mathbf{l}_i$ is the $i$th column of $\mathbf{L}$. Each of the matrices $\mathbf{X}$, $\mathbf{B}$, $\mathbf{F}$, $\mathbf{D}$, and $\mathbf{L}$ is independent of both $\boldsymbol{\beta}$ and $\sigma^2$, so that $\delta_i^2$ depends on $\boldsymbol{\beta}$ and $\sigma^2$ if $\mathbf{X}'\mathbf{B}\mathbf{F}\mathbf{D}^{-\frac{1}{2}}\mathbf{l}_i \neq \mathbf{o}$. If $\lambda_i \neq 0$, then $\delta_i^2$ should not depend on $\sigma^2$ or $\boldsymbol{\beta}$; in this case we call $\mathcal{F}(t)$ *calculable.* For instance, $\mathcal{F}(t)$ is calculable if $\mathbf{B}' = \mathbf{M}$, i.e. $\mathbf{w} = \hat{\mathbf{u}}$.

We wish to apply $\mathbf{w} = \mathbf{B}'\mathbf{y}$ in several test statistics; in particular, we consider $\mathbf{C} = \mathbf{I}$. In that case we have:

$$\begin{aligned} rank(\Lambda_{(r)}) &= rank\,[\mathbf{D}^{\frac{1}{2}}\mathbf{F}'(\mathbf{A} - t\mathbf{C})\mathbf{F}\mathbf{D}^{\frac{1}{2}}] \\ &= rank\,[\mathbf{F}'(\mathbf{A} - t\mathbf{I}_{(p)})\mathbf{F}] \\ &= rank\,(\mathbf{F}'\mathbf{A}\mathbf{F} - t\mathbf{I}_{(r)}) = r \end{aligned}$$

unless $t$ is an eigenvalue of $\mathbf{F}'\mathbf{A}\mathbf{F}$. Hence, for all but at most $r$ distinct values of $t$, all of $\lambda_1, \lambda_2, \ldots, \lambda_r$ are nonzero. Consequently, none of $\delta_1^2, \delta_2^2, \ldots, \delta_r^2$ should depend on $\boldsymbol{\beta}$ or $\sigma^2$. This is equivalent to $\mathbf{L}'\mathbf{D}^{-\frac{1}{2}}\mathbf{F}'\mathbf{B}'\mathbf{X} = \mathbf{0}$, so that $\mathbf{F}'\mathbf{B}'\mathbf{X} = \mathbf{0}$. Since $\mathbf{B}' = \mathbf{F}\mathbf{G}'$ for some $n \times r$ matrix $\mathbf{G}$, we have $\mathbf{F}'\mathbf{B}'\mathbf{X} = \mathbf{F}'\mathbf{F}\mathbf{G}'\mathbf{X} = \mathbf{G}'\mathbf{X} = \mathbf{0}$. But $\mathbf{G}'\mathbf{X} = \mathbf{0}$ if and only if $\mathbf{F}\mathbf{G}'\mathbf{X} = \mathbf{0}$, since the columns of $\mathbf{F}$ are linearly independent. In that

1. Recently L'Esperance et al. (1976) proposed the direct calculation of Imhof's procedure in terms of complex numbers, whereas Koerts and Abrahamse use real numbers. Provided that the computer program is efficient the complex number technique is faster than the real number technique.

case $\mathbf{B}'\mathbf{X} = \mathbf{0}$ and hence $\boldsymbol{\delta} = \mathbf{o}$. Summarizing, we have found that $\mathcal{F}(t)$ is calculable if and only if $\mathbf{B}$ satisfies :

$$\mathbf{B}'\mathbf{X} = \mathbf{0} \tag{2.11}$$

This implies :

$$rank\,(\mathbf{B}) = r \leqslant min(p, n\text{-}k) \tag{2.12}$$

In accordance with (2.10) :

$$\mathcal{F}(t) = Pr[\sum_{i=1}^{r} \lambda_i z_i^2 \leqslant 0] \tag{2.13}$$

where $\lambda_1 \leqslant \lambda_2 \leqslant \ldots \leqslant \lambda_r$ are the eigenvalues of $\mathbf{D}^{1/2}\mathbf{F}'(\mathbf{A} - t\mathbf{C})\mathbf{F}\mathbf{D}^{1/2}$, and $z_1^2, z_2^2, \ldots, z_r^2$ are mutually stochastically independent central $\chi^2(1)$-variables; $\mathbf{F}$ and $\mathbf{D}$ are matrices such that $\mathbf{F}'\mathbf{F} = \mathbf{I}_{(r)}$ and $\mathbf{FDF}' = \mathbf{B}'\boldsymbol{\Gamma}\mathbf{B}$. Sometimes it is convenient to refer to the nonzero $\lambda_i$ as the nonzero eigenvalues of $\mathbf{S}'\mathbf{B}(\mathbf{A} - t\mathbf{C})\mathbf{B}'\mathbf{S}$, where $\mathbf{S}$ is an $n \times n$ matrix satisfying $\mathbf{SS}' = \boldsymbol{\Gamma}$. The fact that the nonzero $\lambda_i$ are the nonzero eigenvalues of $\mathbf{S}'\mathbf{B}(\mathbf{A} - t\mathbf{C})\mathbf{B}'\mathbf{S}$ is proved as follows. From $(\mathbf{FD}^{1/2})(\mathbf{FD}^{1/2})' = (\mathbf{B}'\mathbf{S})(\mathbf{B}'\mathbf{S})'$ and from Result 1.A.6.3 it follows that $\mathbf{B}'\mathbf{S} = \mathbf{FD}^{1/2}\mathbf{G}'$, where $\mathbf{G}$ is a matrix satisfying $\mathbf{G}'\mathbf{G} = \mathbf{I}$. Given $\mathbf{D}^{1/2}\mathbf{F}'(\mathbf{A} - t\mathbf{C})\mathbf{FD}^{1/2} = \mathbf{L}\boldsymbol{\Lambda}\mathbf{L}'$ with $\mathbf{L}' = \mathbf{L}^{-1}$, we have $\mathbf{S}'\mathbf{B}(\mathbf{A} - t\mathbf{C})\mathbf{B}'\mathbf{S} = (\mathbf{GL})\boldsymbol{\Lambda}(\mathbf{GL})'$ with $(\mathbf{GL})'(\mathbf{GL}) = \mathbf{L}'\mathbf{G}'\mathbf{GL} = \mathbf{I}$, so that all nonzero eigenvalues are contained in $\boldsymbol{\Lambda}$.

The probability (2.13) can be calculated by means of the procedures of Imhof and of Pan Jie-jian. It is important to realize that these methods enable us to calculate $\mathcal{F}(t)$ for given $t$; there is no procedure to calculate $t$ for given $\mathcal{F}(t)$. Therefore, repetitive use of (one of) these methods is necessary when one wishes to find $t$ such that $\mathcal{F}(t) = \alpha$, where $\alpha$ is given. Generally, some ten iterations are necessary to find a value $t_0$ such that $|\mathcal{F}(t_0) - \alpha| < 0.0001$. The iteration procedure is described in Section 2.6.

## 2.3. Tabulable distribution functions

When a test is applied, it is good custom to choose the significance level $\alpha$ independent of the outcome of the test statistic. Let the critical region have the form $T \leqslant t$. One chooses $\alpha$, for instance, $\alpha = 0.05$. Then the significance point $t$ is determined such that $\mathcal{F}(t|\mathcal{H}_0) = \alpha$. Let $T_0$ be the outcome of $T$; then $\mathcal{H}_0$ must be rejected if $T_0 \leqslant t$ and $\mathcal{H}_0$ must be accepted if $T_0 > t$. This procedure is commonly used in practice. An alternative approach would be: calculate $\mathcal{F}(T_0|\mathcal{H}_0) = \hat{\alpha}$ the critical level of the test (see Lehmann 1959, p. 62), and reject $\mathcal{H}_0$ if $\hat{\alpha} \leqslant \alpha$, otherwise accept $\mathcal{H}_0$. We refer to this approach in Section 5.3.

The determination of $t$ such that $\mathcal{F}(t|\mathcal{H}_0) = \alpha$ often requires very complicated calculations, which must be performed for each application of the test. This is because the distribution of $\mathbf{w} = \mathbf{B}'\mathbf{y}$, with $\mathbf{B}$ satisfying $\mathbf{B}'\mathbf{X} = \mathbf{0}$, generally depends on $\mathbf{X}$, and hence $T$, being a function of $\mathbf{w}$, generally has a distribution which depends on $\mathbf{X}$. For instance, the distribution function $\mathcal{F}(d^*|\mathcal{H}_0)$ of $d$ in (2.5) is equal to $Pr[\sum_{i=1}^{n-k} \lambda_i z_i^2 \leqslant 0]$, see (2.13), where $\lambda_1 \leqslant \lambda_2 \leqslant \ldots \leqslant \lambda_{n-k}$ are the eigenvalues of $\mathbf{N}'(\mathbf{A}_d - d^*\mathbf{I}_{(n)})\mathbf{N}$, in which $\mathbf{N}$ is an orthogonal $n \times (n-k)$ matrix such that $\mathbf{N}\mathbf{N}' = \mathbf{M}$ (see Section 1.3). Generally, if $\mathbf{X}$ varies, then $\mathbf{N}$ varies, and hence the $\lambda$'s vary. The computational inconvenience of such a test is a serious drawback for applications.

In order to diminish the calculations, two approaches can be followed: either to approximate $t$ or to use a vector $\mathbf{w}$ in $T$ such that the distribution of $T$ does not depend on $\mathbf{X}$. One type of approximation is the so-called *bounds test*, which we discuss in Section 2.7. Another type of approximation is the *point approximation* for $t$. Several point approximation methods are discussed in Durbin and Watson (1971).

In this study, we follow the second approach, i.e. we consider vectors $\mathbf{w}$, $\mathbf{w} = \mathbf{B}'\mathbf{y}$ with $\mathbf{B}'\mathbf{X} = \mathbf{0}$, such that $\mathcal{F}(t|\mathcal{H}_0)$ is independent of the specification of $\mathbf{X}$. In this case we call $\mathcal{F}(t|\mathcal{H}_0)$ *tabulable* and we speak of a *tabulable test*, i.e. for given $\alpha = \mathcal{F}(t|\mathcal{H}_0)$ it is possible to calculate $t$ once and for all, so that it makes sense to tabulate $t$ for several well-chosen values of $\alpha$. This approach was initiated by Theil (1965, 1968) and Koerts (1965), and extended by Abrahamse and Koerts (1971). These authors developed $\mathbf{X}$-independently distributed vectors $\mathbf{w}$ (Theil and Koerts: the *BLUS* vector; Abrahamse and Koerts: the *new estimators*; see Section 3.3). Given $\mathbf{y} \sim \mathcal{N}(\mathbf{X}\boldsymbol{\beta}, \sigma^2\Gamma)$ and $\mathcal{H}_0 : \Gamma = \Gamma_0$, the distribution of $\mathbf{w} = \mathbf{B}'\mathbf{y}$ with $\mathbf{B}'\mathbf{X} = \mathbf{0}$ under $\mathcal{H}_0$ is $\mathcal{N}(\mathbf{o}, \sigma^2\mathbf{B}'\Gamma_0\mathbf{B})$. Therefore, if $\mathbf{B}'\Gamma_0\mathbf{B} = \Omega$, where $\Omega$ is some fixed symmetric nonnegative definite matrix with order $p \times p$ and rank $r$, then the distribution of **w** is $\mathcal{N}(\mathbf{o}, \sigma^2 \Omega)$, which is the same for all specifications of **X**. A *BLUS* vector is constructed such that $\Omega = \mathbf{I}_{(n-k)}$, while the *new estimators* are constructed such that $\Omega$ is a fixed symmetric idempotent matrix with order $n \times n$ and rank $n - k$. Obviously, if $\mathbf{w} \sim \mathcal{N}(\mathbf{o}, \sigma^2 \Omega)$ with $\Omega$ fixed, then the distribution of $T$ does not depend on **X**.

Summarizing, given $T = \mathbf{w}'\mathbf{A}\mathbf{w}/\mathbf{w}'\mathbf{C}\mathbf{w}$ and $\mathcal{H}_0 : \Gamma = \Gamma_0$, then $\mathcal{F}(t|\mathcal{H}_0)$ is tabulable if:

$$\begin{array}{ll} \mathbf{w} = \mathbf{B}'\mathbf{y} & \mathbf{B} \text{ has order } n \times p \text{ and rank } r \\ \mathbf{B}'\mathbf{X} = \mathbf{0} & r \leqslant min\,(n - k, p) \\ \mathbf{B}'\Gamma_0\mathbf{B} = \Omega = \mathbf{K}\mathbf{K}' & \Omega \text{ is a fixed symmetric non-} \\ & \text{negative definite } p \times p \text{ matrix} \\ & \text{with rank } r, \mathbf{K} \text{ has order } p \times r. \end{array} \tag{2.14}$$

We have:

$$\mathcal{F}(t|\mathcal{H}_0) = Pr[\sum_{i=1}^{r} \lambda_i z_i^2 \leqslant 0] \tag{2.15}$$

where $\lambda_1 \leqslant \lambda_2 \leqslant \ldots \leqslant \lambda_r$ are the eigenvalues of $\mathbf{K}'(\mathbf{A} - t\mathbf{C})\mathbf{K}$ and $z_1^2, z_2^2, \ldots, z_r^2$ are mutually independent central $\chi^2(1)$-variables. We call $\sigma^2\Omega$ the *a priori fixed* (namely prior to the specification of **X**) *covariance matrix* of **w**. Note that one is free to choose the order of $\Omega$, but the choice of the rank of $\Omega$ is restricted by the order of **X**, namely $rank(\Omega) \leqslant n$-$k$. Given the $p \times p$ matrix $\Omega$ with rank $r$, the $p \times r$ matrix **K** such that $\mathbf{K}\mathbf{K}' = \Omega$ is not unique. **K** may be replaced by $\mathbf{K}_t$, $\mathbf{K}_t = \mathbf{K}\mathbf{T}$ where **T** is an arbitrary orthogonal $r \times r$ matrix (see Result 1.A.6.3). Clearly, $\mathbf{K}'(\mathbf{A} - t\mathbf{C})\mathbf{K}$ and $\mathbf{K}_t'(\mathbf{A} - t\mathbf{C})\mathbf{K}_t$ have the same eigenvalues.

## 2.4. On the choice of w and Ω

The entire class of vectors **w** satisfying (2.14) is defined in Section 3.2. There it is found that:

$$\mathbf{w} = \mathbf{B}'\mathbf{y} = \mathbf{K}\mathbf{H}'\bar{\mathbf{N}}'\mathbf{S}^{-1}\mathbf{y} \tag{2.16}$$

where **K** satisfies $\mathbf{K}\mathbf{K}' = \Omega$, **S** is an $n \times n$ matrix satisfying $\mathbf{S}\mathbf{S}' = \Gamma_0$,

$\bar{\mathbf{N}}$ is an orthogonal $n \times (n\text{-}k)$ matrix such that $\mathcal{M}(\bar{\mathbf{N}}) = \mathcal{M}(\mathbf{S}^{-1}\mathbf{X})^{\perp}$ (see Section 1.3) and **H** is an arbitrary orthogonal $(n\text{-}k) \times r$ matrix. We wish to have a vector **w** such that the test, to which **w** is applied, is as powerful as possible. Given $\Gamma_0$ and **X**, the matrix **S** can be determined from $\Gamma_0$, for instance, $\mathbf{S} = \Gamma_0^{1/2}$, and $\bar{\mathbf{N}}$ can be determined from **S** and **X** (both **S** and $\bar{\mathbf{N}}$ are not unique). Having **S** and $\bar{\mathbf{N}}$ determined, we are still free to choose **K** and **H**.

With respect to the choice of **w** (i.e. the choice of **K** and **H** for given **S** and $\bar{\mathbf{N}}$) we adopt the following two-step approach. In the first place, $\Omega$ is regarded as given and **H** is chosen such that **w** is "as close as possible" to $\mathbf{J}'\mathbf{u}$, where **J** is an arbitrary $n \times p$ matrix. For instance, $\mathbf{J} = \mathbf{I}_{(n)}$ or $\mathbf{J} = \mathbf{M}^*$. This problem is formalized and solved in Chapter 3. In the second place, we choose $\Omega$, which is the subject of Chapter 4. There we restrict our attention to practical testing problems with $\mathcal{H}_0 : \Gamma = \mathbf{I}$, so that $\mathbf{u}^* = \hat{\mathbf{u}} = \mathbf{My} \sim \mathcal{N}(\mathbf{o}, \sigma^2\mathbf{M})$. From a set of $n \times k$ **X**-matrices we determine a *mean* **X**, and we put $\Omega$ equal to the **M**-matrix corresponding to this *mean* **X**.

The least-squares residuals play an important role in the most powerful tests, as set out in Section 2.1. However, such tests are not tabulable. We may hope that the tabulable tests, based on **w** in accordance with the above two-step least-squares approximation procedure (**w** close to **u** or $\hat{\mathbf{u}}$ for given $\Omega$, and $\Omega$ equal to the *mean* **M**) are close to most powerful.

The power of the test at significance level $\alpha$ for $\mathcal{H}_0 : \Gamma = \mathbf{I}$ against $\mathcal{H}_A : \Gamma = \Gamma_A$, defined by the critical region $T = \mathbf{w}'\mathbf{A}\mathbf{w}/\mathbf{w}'\mathbf{C}\mathbf{w} \leqslant t$ is:

$$\mathcal{P}(t|\mathcal{H}_A) = Pr[\sum_{i=1}^{r} \nu_i z_i^2 \leqslant 0] \qquad (2.17)$$

where $z_1^2, z_2^2, \ldots, z_r^2$ are mutually stochastically independent central $\chi^2(1)$-variables and $\nu_1 \leqslant \nu_2 \leqslant \ldots \leqslant \nu_r$ are the eigenvalues of $\mathbf{D}^{1/2}\mathbf{F}'(\mathbf{A} - t\mathbf{C})\mathbf{F}\mathbf{D}^{1/2}$, **F** and **D** satisfying $\mathbf{B}'\Gamma_A\mathbf{B} = \mathbf{FDF}'$ and $\mathbf{F}'\mathbf{F} = \mathbf{I}_{(r)}$; see (2.13). Note that $\mathcal{H}_0 : \Gamma = \mathbf{I}$ implies that we may take $\mathbf{S} = \mathbf{I}$, so that $\mathbf{B}' = \mathbf{KH}'\mathbf{N}'$, where **N** is an orthogonal $n \times (n\text{-}k)$ matrix such that $\mathcal{M}(\mathbf{N}) = \mathcal{M}(\mathbf{X})^{\perp}$. Hence, $\mathbf{B}'\Gamma_0\mathbf{B} = \mathbf{KK}'$ and $\mathbf{B}'\Gamma_A\mathbf{B} = \mathbf{KH}'\mathbf{N}'\Gamma_A\mathbf{NHK}'$. Given $\alpha$, **A**, and **C**, the significance point $t$ depends on **K** alone, while the power depends on **K** (which represents the choice of $\Omega$), and on **H** (which represents the selection of **w** for given $\Omega$), and on **N** (which represents the **X**-matrix), and on $\Gamma_A$ (which represents the alternative hypothesis). In Chapter 5 several vectors **w**, i.e. $\mathbf{w}_1 = \mathbf{K}_1\mathbf{H}_1'\mathbf{N}_1'\mathbf{y}$, $\mathbf{w}_2 = \mathbf{K}_2\mathbf{H}_2'\mathbf{N}_2'\mathbf{y}$, and so on, are evaluated by means of the power scores in three specific tests, to be described in Section 2.5. In all cases we take $\mathcal{H}_0 : \Gamma = \mathbf{I}$.

## 2.5 Three specific tests

We describe three tests using test statistics of the type $T$; see (2.7). Since $T$ is a function of $\mathbf{w}$, which vector is not completely specified at this stage, each of the three tests in fact represents a set of tests. The test with test statistic $T$ will be denoted by: *test(T)*. *Test(Q)* is a test (set of tests) against positive autocorrelation, *test(S)* and *test (V)* are tests against heterovariance.

*Autocorrelation test (Q)*
Given $\mathbf{y} \sim \mathcal{N}(\mathbf{X}\boldsymbol{\beta}, \sigma^2\Gamma)$, with $\Gamma^{-1}$ given in (2.3), we consider the test for $\mathcal{H}_0 : \rho = 0$ against $\mathcal{H}_A : \rho > 0$ with the critical region of the form:

$$Q = \frac{\mathbf{w}'\mathbf{A}_d\mathbf{w}}{\mathbf{w}'\mathbf{E}\mathbf{w}} \leqslant q \qquad (2.18)$$

where the matrices $\mathbf{A}_d$, see (2.4), and $\mathbf{E}$, see Section 1.A.1, have order $p \times p$, In Chapter 5 the powers are calculated for $\mathcal{H}_A : \rho = 0.8$. When we take $\mathbf{w}$ equal to $\hat{\mathbf{u}}$, so that $p = n$, and if $\mathbf{X}$ contains the constant term vector $\boldsymbol{\iota}$, then $\mathbf{M}\boldsymbol{\iota} = \mathbf{o}$ in view of $\mathbf{MX} = \mathbf{0}$, and hence $\mathbf{EM} = \mathbf{M}$, so that $\mathbf{w}'\mathbf{E}\mathbf{w} = \hat{\mathbf{u}}'\mathbf{E}\hat{\mathbf{u}} = \mathbf{y}'\mathbf{MEMy} = \mathbf{y}'\mathbf{MMy} = \hat{\mathbf{u}}'\hat{\mathbf{u}}$. In this case it is seen that $Q = d$; see (2.5). It may be noted that, with respect to *test(d)*, Durbin and Watson consider only $\mathbf{X}$-matrices containing a constant term. In Chapter 5 we also deal with such $\mathbf{X}$-matrices only. The reason why we insert $\mathbf{E}$ into the denominator of $Q$ is, that the *BLUS* test against autocorrelation is defined with $\mathbf{E}$ inclusive (see Koerts and Abrahamse 1969, p. 74). One of the $\mathbf{w}$-vectors in Chapter 5 is a *BLUS* vector.

*Heterovariance test (S)*
Given $\mathbf{y} \sim \mathcal{N}(\mathbf{X}\boldsymbol{\beta}, \sigma^2\Gamma)$ with:

$$\Gamma = \begin{bmatrix} \frac{n}{n-\eta} & 0 & 0 & \cdots & 0 \\ 0 & \frac{n}{n-2\eta} & 0 & \cdots & 0 \\ \vdots & & & & \vdots \\ 0 & 0 & 0 & \cdots & \frac{n}{n-n\eta} \end{bmatrix}$$

we consider the test for $\mathcal{H}_0 : \eta = 0$ against $\mathcal{H}_A : \eta > 0$ with the critical region of the form:

$$S = \frac{\mathbf{w}'\mathbf{A}_s\mathbf{w}}{\mathbf{w}'\mathbf{w}} \geqslant s \tag{2.19}$$

where

$$\mathbf{A}_s = \begin{bmatrix} \frac{1}{p} & 0 & 0 & \cdots & 0 \\ 0 & \frac{2}{p} & 0 & \cdots & 0 \\ \vdots & & & & \vdots \\ 0 & 0 & 0 & \cdots & \frac{p}{p} \end{bmatrix}$$

In Chapter 5 the powers are calculated for $\mathcal{H}_A : \eta = 0.9$. It is seen that $\Gamma^{-1} = \mathbf{I} - \eta\mathbf{A}_s$ for $p = n$. In accordance with (2.2), the test with the critical region of the form:

$$S_1 = \frac{\hat{\mathbf{u}}'(-\mathbf{A}_s)\hat{\mathbf{u}}}{\hat{\mathbf{u}}'\hat{\mathbf{u}}} \leqslant s_1$$

is *UMPS* if the columns of $\mathbf{X}$ are linear combinations of $k$ eigenvectors of $\mathbf{A}_s$. The matrix of eigenvectors of $\mathbf{A}_s$ is $\mathbf{I}_{(n)}$.

We constructed this heterovariance test with two reasons in mind. In the first place, we wished to have gradually increasing variances in such a way that the heterovariance scheme has a realistic significance in economic time series. In the second place, we wished to have a *UMPS* test. Of course, *test(S)* with $\mathbf{w} = \hat{\mathbf{u}}$ is *UMPS* for a restricted class of $\mathbf{X}$-matrices. Every $n \times k$ matrix $\mathbf{X}$ in this class has $n - k$ zero rows and the remaining $k$ rows together form a nonsingular $k \times k$ matrix. Such $\mathbf{X}$-matrices are not very likely in economic time series. However, the fact that the *UMPS* class is unrealistic does not disqualify the test beforehand.

*Heterovariance test (V)*
Given $y \sim \mathcal{N}(X\boldsymbol{\beta}, \sigma^2\Gamma)$ with:

$$\Gamma = \begin{bmatrix} 1^\gamma & 0 & 0 & \dots & 0 \\ 0 & 2^\gamma & 0 & \dots & 0 \\ \vdots & & & & \vdots \\ 0 & 0 & 0 & \dots & n^\gamma \end{bmatrix}$$

we consider the test for $\mathcal{H}_0 : \gamma = 0$ against $\mathcal{H}_A : \gamma > 0$ with the critical region of the form:

$$V = \frac{\mathbf{w}'\mathbf{A}_\nu\mathbf{w}}{\mathbf{w}'(\mathbf{I}_{(p)} - \mathbf{A}_\nu)\mathbf{w}} \leqslant \nu \tag{2.20}$$

where:

$$\mathbf{A}_\nu = \left[\begin{array}{c:c} \mathbf{I}_{(m)} & \mathbf{0} \\ \hdashline \mathbf{0} & \mathbf{0}_{(p-m)} \end{array}\right] \qquad \begin{array}{l} m = \tfrac{1}{2}p \text{ if } p \text{ is even} \\ \\ m = \tfrac{1}{2}p - \tfrac{1}{2} \text{ if } p \text{ is odd} \end{array} \tag{2.21}$$

In Chapter 5 the powers are calculated for $\mathcal{H}_A : \gamma = 0.83$ (the choice of this value of $\gamma$ is related to the choice of $\eta = 0.9$; see Section 5.5). The matrix $\Gamma$ originates from a time series model by Geary (1966). The test statistic $V$, with $\mathbf{w}$ equal to a *BLUS* vector, has been proposed by Theil (1968). Theil suggests adopting that *BLUS* vector, which does not estimate the $k$ middle disturbances (see Section 3.3). If $\mathbf{w}$ is a *BLUS* vector, then $\mathbf{w} \sim \mathcal{N}(\mathbf{o}, \sigma^2\mathbf{I}_{(p)})$ with $p = n - k$. In that case $(p-m)V/m = (\sum_{i=1}^{m} w_i^2/m) / [\sum_{i=m+1}^{p} w_i^2/(p-m)]$ has a central F-(Snedecor) distribution under $\mathcal{H}_0$. Abrahamse (1970) combined $\Gamma$ (from Geary) and $V$ (from Theil) into the above test, taking a particular vector $\mathbf{w}$, which we identify by *hyp* in Chapter 5. Geary computed the efficiency (in terms of variances) of the *o.l.s.* versus the *g.l.s.* $\boldsymbol{\beta}$-estimator for several values of $\gamma$. Abrahamse compared the powers of his test and the efficiencies computed by Geary.

It may be noted that the above test with $\mathbf{w} = \hat{\mathbf{u}}$ is *UMPS* (for a restricted class of **X**-matrices) if $\Gamma$ would be replaced by $\Gamma = \mathbf{I}_{(n)} - \gamma\mathbf{A}_v/(1+\gamma)$ so that $\Gamma^{-1} = \mathbf{I} + \gamma\mathbf{A}_v$. In accordance with (2.2), the test statistic would read $\hat{\mathbf{u}}'\mathbf{A}_v\hat{\mathbf{u}}/\hat{\mathbf{u}}'\hat{\mathbf{u}} = V^*$, and there is a one-to-one correspondence between $V^*$ and $V$ for $0 < V^* < 1$, namely $V - V^* = VV^*$. Therefore, the use of either $V^*$ or $V$ is optional. In Section 5.5, where the various **w**-vectors are evaluated by means of the powers of *test(S)* and *test(V)*, we also evaluate the test statistics $S$ and $V$ by means of the powers of mixed tests, i.e. the alternative hypotheses in *test(S)* and *test(V)* are interchanged.

## 2.6 Significance point calculation

In accordance with (2.15):

$$\mathcal{F}(t|\mathcal{H}_0) = Pr\,[\sum_{i=1}^{r} \lambda_i(t) z_i^2 \leqslant 0]$$

where $\lambda_1(t) \leqslant \lambda_2(t) \leqslant \ldots \leqslant \lambda_r(t)$ are the eigenvalues of $\mathbf{K}'(\mathbf{A} - t\mathbf{C})\mathbf{K}$, $\mathbf{K}$ satisfying $\mathbf{KK}' = \Omega$. The notation $\lambda_i(t)$ expresses the fact that $\lambda_i$ is a function of $t$. Hence, $\lambda_i(0)$ is an eigenvalue of $\mathbf{K}'\mathbf{AK}$. In all our applications $\Omega$ is taken idempotent (see Section 4.2). Hence $\mathbf{K}'\mathbf{K} = \mathbf{I}_{(r)}$. Let $\mathbf{P}$ be a $p \times (p\text{-}r)$ matrix such that the matrix $[\mathbf{P} \vdots \mathbf{K}]$ is square and orthogonal. Then $\mathbf{P}'\mathbf{K} = \mathbf{0}$, $\mathbf{P}'\mathbf{P} = \mathbf{I}_{(p-r)}$, and $\Omega = \mathbf{I} - \mathbf{PP}'$. Usually $p = n$ and $r = n\text{-}k$, so that $\mathbf{P}$ has order $n \times k$, the same as $\mathbf{X}$. In the philosophy of Section 2.4, where we proposed to take $\Omega$ equal to a *mean* **M**, the matrix **P** is to be regarded as an orthogonal *mean* **X**. **P** is usually smaller (fewer columns) than **K**, which makes it attractive to use **P** rather than **K** in calculations, if possible.

Below we establish simple relations between $\lambda_i(t)$ and $\lambda_i(0)$ for each of the three tests described in the previous section, like, for instance, $\lambda_i(t) = \lambda_i(0) - t$ for $i = 1, 2, \ldots, r$. Significance point calculation then requires that $\lambda_i(0)$ be determined once, and that the established relation in each of the iterations be applied. Finally, a general iteration procedure is presented.

*The* $\lambda_i\,(q)$

In the autocorrelation test, $\mathbf{A} - t\mathbf{C}$ becomes $\mathbf{A}_d - q\mathbf{E}$. The matrix $\mathbf{E}$ is idempotent, one eigenvalue is zero and all remaining eigenvalues are equal to 1. The eigenvector corresponding to the zero eigenvalue is $(\boldsymbol{\iota}'\boldsymbol{\iota})^{-½}\boldsymbol{\iota}$, since $\mathbf{E}\boldsymbol{\iota} = \mathbf{o}$. Hence, each square orthogonal matrix with $(\boldsymbol{\iota}'\boldsymbol{\iota})^{-½}\boldsymbol{\iota}$ as one of its columns diagonalizes **E**. The eigenvalues $d_i$ with corresponding eigenvectors $\mathbf{h}_i^*$ of the $p \times p$ matrix $\mathbf{A}_d$ are given by Von Neumann (1941):

$$d_i = 2 - 2\,cos[\pi(i\text{-}1)/p] \qquad i = 1, 2, \ldots, p \qquad (2.22)$$

so that $0 = d_1 < d_2 < \ldots < d_p < 4$; while $\mathbf{h}_1^* = (\boldsymbol{\iota}'\boldsymbol{\iota})^{-½}\boldsymbol{\iota}$ and $h_i^*(j)$, the $j$th element of $\mathbf{h}_i^*$, is:

$$h_i^*(j) = (\frac{p}{2})^{-½} cos[\pi\,(i\text{-}1)\,(j\text{-}½)/p] \qquad \begin{array}{l} j = 1,2, \ldots, p \\ i = 2,3, \ldots, p \end{array} \qquad (2.23)$$

Hence:

$$\mathbf{A}_d - q\mathbf{E} = \mathbf{H}^* \begin{bmatrix} 0 & 0 & 0 & \cdots & 0 \\ 0 & d_2-q & 0 & \cdots & 0 \\ 0 & 0 & d_3-q & \cdots & 0 \\ \vdots & & & & \vdots \\ 0 & 0 & 0 & \cdots & d_p-q \end{bmatrix} \mathbf{H}^{*\prime} \qquad (2.24)$$

We consider some specifications of **P**. In the first place we consider $\mathbf{w} = \hat{\mathbf{u}}$, so that $\mathcal{M}(\mathbf{P}) = \mathcal{M}(\mathbf{X})$; in the second place we consider the case that **w** is a *BLUS* vector, so that $\Omega = \mathbf{I}_{(n\text{-}k)}$ and **P** does not exist; in the third place we consider $\mathbf{P} = \mathbf{h}_1^*$ and **P**-matrices consisting of more than one **h***-column.

When $\mathbf{w} = \hat{\mathbf{u}} = \mathbf{My}$ and **X** contains $\boldsymbol{\iota}$, then $\mathbf{M}\boldsymbol{\iota} = \mathbf{o}$ and $\mathbf{KK}' = \mathbf{M}$, so that $\mathbf{K}'\boldsymbol{\iota} = \mathbf{o}$. Hence, $\mathbf{K}'\mathbf{EK} = \mathbf{I}_{(n\text{-}k)}$ and $\mathbf{K}'(\mathbf{A}_d - q\mathbf{E})\mathbf{K} = \mathbf{K}'\mathbf{A}_d\mathbf{K} - q\mathbf{I}_{(n\text{-}k)}$. It follows that the $\lambda_i(0)$ are the eigenvalues of $\mathbf{K}'\mathbf{A}_d\mathbf{K}$ and:

$$\lambda_i(q) = \lambda_i(0) - q \qquad i = 1,2, \ldots, n\text{-}k$$

When $\Omega = \mathbf{I}_{(p)}$, then we may take **K** equal to $\mathbf{I}_{(p)}$, and we have $\mathbf{K}'(\mathbf{A}_d - q\mathbf{E})\mathbf{K} = \mathbf{A}_d - q\mathbf{E}$. It follows that $\lambda_i(0) = d_i$, see (2.22), and:

$$\lambda_1(q) = 0$$
$$\lambda_i(q) = \lambda_i(0) - q \qquad i = 2,3, \ldots, p$$

where $p = n\text{-}k$ in the case that **w** is a *BLUS* vector. When $\mathbf{P} = \mathbf{h}_1^*$ we have $\Omega = \mathbf{I}_{(p)} - \mathbf{PP}' = \mathbf{E}_{(p)}$. Taking **K** such that $[\mathbf{P} \vdots \mathbf{K}] = \mathbf{H}^*$, we have $\mathbf{K}'\mathbf{E}_{(p)}\mathbf{K} = \mathbf{I}_{(p-1)}$ and $\mathbf{K}'\mathbf{A}_d\mathbf{K}$ is a diagonal $(p-1) \times (p-1)$ matrix with its $i$th diagonal element equal to $d_{i+1}$, see (2.22). It follows that the nonzero eigenvalues of $\mathbf{K}'(\mathbf{A}_d - q\mathbf{E})\mathbf{K}$ in the case that $\Omega = \mathbf{E}_{(p)}$ are equal to those in the case that $\Omega = \mathbf{I}_{(p)}$. When **P** consists of $\mathbf{h}_1^*$ and one or more other $\mathbf{h}^*$-columns, so that **K** consists of $\mathbf{h}_{j1}^*, \mathbf{h}_{j2}^*, \ldots, \mathbf{h}_{jr}^*$, say; then $\mathbf{K}'\mathbf{E}_{(p)}\mathbf{K} = \mathbf{I}_{(r)}$ and $\mathbf{K}'\mathbf{A}_d\mathbf{K}$ is diagonal and contains $d_{j1}, d_{j2}, \ldots, d_{jr}$ on its diagonal; see (2.22). Hence, $\lambda_i(0) = d_{ji}$ for $ji = j1, j2, \ldots, jr$, and:

$$\lambda_i(q) = \lambda_i(0) - q \qquad i = 1,2, \ldots, r$$

However, if $j1 = 1$, so that **P** does not contain $\mathbf{h}_1^*$, then $\mathbf{K}'\mathbf{EK}$ has leading element zero, so that $\lambda_1(q) = 0 \neq \lambda_1(0) - q$. For instance, let $\mathbf{P} = [\mathbf{h}_2^* \vdots \mathbf{h}_4^*]$ be a $p \times 2$ matrix, so that $\mathbf{h}_1^*, \mathbf{h}_3^*, \mathbf{h}_5^*, \mathbf{h}_6^*, \ldots, \mathbf{h}_p^*$ are the $r = p - 2$ columns of the corresponding **K**-matrix, then the eigenvalues of $\mathbf{K}'(\mathbf{A}_d - q\mathbf{E})\mathbf{K}$ are:

$$\lambda_1(q) = 0;\ \lambda_2(q) = d_3 - q;\ \lambda_3(q) = d_5 - q;\ \lambda_4(q) = d_6 - q;\ \ldots;$$
$$\lambda_r(q) = d_p - q.$$

Analogously, when $\mathbf{P} = [\mathbf{h}_1^* \vdots \mathbf{h}_2^* \vdots \mathbf{h}_4^*]$, then we find:

$$\lambda_1(q) = d_3 - q;\ \lambda_2(q) = d_5 - q;\ \lambda_3(q) = d_6 - q;\ \ldots;\ \lambda_r(q) = d_p - q$$

where $r = p - 3$. We see that the nonzero eigenvalues are the same for $\mathbf{P} = [\mathbf{h}_2^* \vdots \mathbf{h}_4^*]$ and $\mathbf{P} = [\mathbf{h}_1^* \vdots \mathbf{h}_2^* \vdots \mathbf{h}_4^*]$ i.e. $\mathbf{h}_1^*$ can freely be transferred from **K** to **P**, which fact is due to $\mathbf{A}_d\mathbf{h}_1^* = \mathbf{E}\mathbf{h}_1^* = \mathbf{o}$. This fact has already been illustrated above, in the cases where $\Omega = \mathbf{I}_{(p)}$ and $\Omega = \mathbf{E}_{(p)}$.

*The $\lambda_i(s)$*

In the first heterovariance test, $\mathbf{A} - t\mathbf{C}$ becomes $\mathbf{A}_s - s\mathbf{I}$. The $\lambda_i(s)$ are the eigenvalues of $\mathbf{K}'(\mathbf{A}_s - s\mathbf{I}_{(p)})\mathbf{K} = \mathbf{K}'\mathbf{A}_s\mathbf{K} - s\mathbf{I}_{(r)}$. Hence, the $\lambda_i(0)$ are the eigenvalues of $\mathbf{K}'\mathbf{A}_s\mathbf{K}$ and:

$$\lambda_i(s) = \lambda_i(0) - s \qquad i = 1, 2, \ldots, r$$

The $\lambda_i(0)$ can be determined in a simple way if $\Omega$ is diagonal. Since $\Omega$ is idempotent, its diagonal contains $r$ elements equal to 1, all other elements zero, and $\mathbf{K}$ can be taken equal to a $p \times r$ matrix, whose columns are $r$ different columns from $\mathbf{I}_{(p)}$, say $\mathbf{e}_{j1}, \mathbf{e}_{j2}, \ldots, \mathbf{e}_{jr}$, with $j1 < j2 < \ldots < jr$. In this case we have $0 < \lambda_1(0) = \frac{j1}{p} < \lambda_2(0) = \frac{j2}{p} < \ldots < \lambda_r(0) = \frac{jr}{p} \leqslant 1$.

*The $\lambda_i(\nu)$*

In the second heterovariance test, $\mathbf{A} - t\mathbf{C}$ becomes $\mathbf{A}_\nu - \nu(\mathbf{I}_{(p)} - \mathbf{A}_\nu) = (1+\nu)\mathbf{A}_\nu - \nu\mathbf{I}_{(p)}$. The $\lambda_i(\nu)$ are the eigenvalues of $\mathbf{K}'[(1+\nu)\mathbf{A}_\nu - \nu\mathbf{I}_{(p)}]\mathbf{K} = (1+\nu)\mathbf{K}'\mathbf{A}_\nu\mathbf{K} - \nu\mathbf{I}_{(r)}$. Hence, the $\lambda_i(0)$ are the eigenvalues of $\mathbf{K}'\mathbf{A}_\nu\mathbf{K}$ and:

$$\lambda_i(\nu) = (1+\nu)\,\lambda_i(0) - \nu \qquad i = 1, 2, \ldots, r$$

When $\Omega$ is diagonal then $\mathbf{K}$ is a matrix consisting of $\mathbf{e}_{j1}, \mathbf{e}_{j2}, \ldots, \mathbf{e}_{jr}$; see the preceding subsection. Let $t$ of the $r$ indices $j1, j2, \ldots, jr$ be smaller than or equal to $m$; see (2.21) for $m$, then $\lambda_i(0) = 1$ for $i = 1, 2, \ldots, t$ and $\lambda_i(0) = 0$ for $i = t+1, t+2, \ldots, r$.

For instance, if $\Omega = \mathbf{I}_{(p)}$ then $r = p$ and $t = m$. If $\mathbf{P} = \mathbf{e}_1$, so that $\Omega = \mathbf{I}_{(p)} - \mathbf{e}_1\mathbf{e}_1'$ then $r = p-1$ and $t = m-1$. In particular, if $n = 15$ and $\mathbf{P} = [\mathbf{e}_7 \vdots \mathbf{e}_8 \vdots \mathbf{e}_9]$, so that the first six and the last six diagonal elements of $\Omega$ are equal to 1 and all other elements zero, then $r = 12$ and $t = 6$. The cases $\Omega = \mathbf{I}_{(12)}$ and $\mathbf{P} = [\mathbf{e}_7 \vdots \mathbf{e}_8 \vdots \mathbf{e}_9]$ are further compared at the end of Section 5.1.

For arbitrary idempotent $\Omega$-matrices, there is a simple relation between the eigenvalues of $\mathbf{K}'\mathbf{A}_\nu\mathbf{K}$ and the eigenvalues of $\mathbf{P}'\mathbf{A}_\nu\mathbf{P}$, as Abrahamse (1970) showed. Usually, the order of $\mathbf{P}'\mathbf{A}_\nu\mathbf{P}$ is much smaller than the order of $\mathbf{K}'\mathbf{A}_\nu\mathbf{K}$, so that it is very advantageous to calculate the eigenvalues of $\mathbf{P}'\mathbf{A}_\nu\mathbf{P}$ and then to apply the relation.

A generalization of the relation reads as follows: let $s$ be the number of nonzero eigenvalues $d_i$ of $\mathbf{P}'\mathbf{A}_\nu\mathbf{P}$ with $0 < d_1 \leqslant d_2 \leqslant \ldots \leqslant d_s$; let:

$$\begin{aligned} \delta_i &= 1 && \text{for } i = 1, 2, \ldots, m-s \\ \delta_{m-s+1} &= 1-d_i && \text{for } i = 1, 2, \ldots, s \\ \delta_{m+i} &= 0 && \text{for } i = 1, 2, \ldots, p-m \end{aligned}$$

then:

$$\lambda_i(0) = \delta_{r+1-i} \qquad \text{for } i = 1, 2, \ldots, r$$

First we remark that $s \leqslant m = rank(\mathbf{A}_\nu)$ and that $d_s \leqslant 1$, in accordance with Lemma 2 in Section 2.7. Using $\mathbf{A}_\nu = \mathbf{A}_\nu' = \mathbf{A}_\nu^2$, the nonzero eigenvalues of $\mathbf{P}'\mathbf{A}_\nu\mathbf{P}$ are equal to those of $\mathbf{A}_\nu\mathbf{PP}'\mathbf{A}_\nu$ (see Result 1.A.6.4). We may write $\mathbf{A}_\nu\mathbf{PP}'\mathbf{A}_\nu = \mathbf{HDH}'$, where $\mathbf{D}$ is an $m \times m$ matrix whose last $s$ diagonal elements are $d_1, d_2, \ldots, d_s$, all other elements zero and $\mathbf{H}$ is a $p \times m$ matrix satisfying $\mathbf{H}'\mathbf{H} = \mathbf{I}_{(m)}$. The last $s$ columns of $\mathbf{H}$ span the $s$-dimensional subspace $\mathcal{M}(\mathbf{A}_\nu\mathbf{P})$ of the $m$-dimensional space $\mathcal{M}(\mathbf{A}_\nu)$. The first $m$-$s$ columns of $\mathbf{H}$ can always be chosen such that $\mathbf{H}$ spans $\mathcal{M}(\mathbf{A}_\nu)$. Then $\mathbf{A}_\nu = \mathbf{HG}'$ for some $p \times m$ matrix $\mathbf{G}$ with rank $m$. Now $\mathbf{A}_\nu = \mathbf{A}_\nu'\mathbf{A}_\nu$ implies $\mathbf{HG}' = \mathbf{GH}'\mathbf{HG}' = \mathbf{GG}'$, and in view of the fact that $(\mathbf{G}'\mathbf{G})^{-1}$ exists, we find $\mathbf{H} = \mathbf{G}$, so that $\mathbf{A}_\nu = \mathbf{HH}'$. Then $\mathbf{A}_\nu\mathbf{KK}'\mathbf{A}_\nu = \mathbf{A}_\nu - \mathbf{A}_\nu\mathbf{PP}'\mathbf{A}_\nu = \mathbf{HH}' - \mathbf{HDH}' = \mathbf{H}(\mathbf{I} - \mathbf{D})\mathbf{H}'$. Since the nonzero eigenvalues of $\mathbf{K}'\mathbf{A}_\nu\mathbf{K}$ are equal to those of $\mathbf{A}_\nu\mathbf{KK}'\mathbf{A}_\nu$, all (at most $m$) nonzero eigenvalues of $\mathbf{K}'\mathbf{A}_\nu\mathbf{K}$ are found on the diagonal of $\mathbf{I} - \mathbf{D} = \Delta$. The diagonal elements $\delta_1$, $\delta_2, \ldots, \delta_m$ are extended by zeros, in order to account for the case $r > m$, which is the usual case.

*An iteration procedure*

Our procedure for the calculation of significance points runs as follows. Let $\alpha$ be the significance level, so that we seek $t_\alpha$ such that $\alpha$-$\epsilon < \mathcal{F}(t_\alpha) < \alpha$+$\epsilon$, where $\epsilon$ indicates the desired accuracy. Determine $t_A$ and $t_B$ such that:

$$\mathcal{F}(t_A) < \alpha < \mathcal{F}(t_B)$$

The choice of $t_A$ and $t_B$ in the three specific tests is considered below. Take $t = (t_A + t_B)/2$ and calculate $\mathcal{F}(t)$. Then either $\alpha$-$\epsilon < \mathcal{F}(t) < \alpha$+$\epsilon$ or $\mathcal{F}(t) \leqslant \alpha$-$\epsilon$ or $\mathcal{F}(t) \geqslant \alpha$+$\epsilon$. If $\alpha$-$\epsilon < \mathcal{F}(t) < \alpha$+$\epsilon$, put $t_\alpha$ equal to $t$ and the procedure is finished. If $\mathcal{F}(t) \leqslant \alpha$-$\epsilon$, put $\mathcal{F}(t_A)$ equal to $\mathcal{F}(t)$ and $t_A$ equal to $t$, while $\mathcal{F}(t_B)$ and $t_B$ remain unchanged. If $\mathcal{F}(t) \geqslant \alpha$+$\epsilon$, put $\mathcal{F}(t_B)$ equal to $\mathcal{F}(t)$ and $t_B$ equal to $t$, while $\mathcal{F}(t_A)$ and $t_A$ remain unchanged. If the procedure is not finished, again take $t = (t_A + t_B)/2$, where either $t_A$ or $t_B$ has a new value, calculate $\mathcal{F}(t)$, and so on, yielding either a new $\mathcal{F}(t_A)$ and $t_A$, or a new $\mathcal{F}(t_B)$ and $t_B$, or $t_\alpha$. Each time that $\mathcal{F}(t)$ does not lie between $\alpha$-$\epsilon$ and $\alpha$+$\epsilon$, a new iteration must be performed. After five iterations, the simple interpolation $t = (t_A + t_B)/2$ is replaced by the linear interpolation $t = t_A + (t_B - t_A)[\alpha - \mathcal{F}(t_A)] / [\mathcal{F}(t_B) - \mathcal{F}(t_A)]$. After the first five iterations, the length of the interval $[t_A, t_B]$ is 1/32 of the length of the initial interval $[t_A, t_B]$, which is usually small enough for a linear interpolation to be advantageous.

In the case of *test(S)*, $\alpha$ must be replaced by 1-$\alpha$: then the procedure finds $s_\alpha$ such that $1\text{-}\alpha\text{-}\epsilon < \mathcal{F}(s_\alpha) < 1\text{-}\alpha\text{+}\epsilon$, or equivalently, $\alpha\text{-}\epsilon < Pr[S > s_\alpha] < \alpha\text{+}\epsilon$.

The initial values of $t_A$ and $t_B$ may be chosen such that $\mathcal{F}(t_A) = 0$ and $\mathcal{F}(t_B) = 1$. Note that $\mathcal{F}(t_A) = 0$ if $\mathbf{A} - t_A\mathbf{C}$ is nonnegative definite, in brief, $\mathbf{A} - t_A\mathbf{C} \geqslant \mathbf{0}$; analogously, $\mathcal{F}(t_B) = 1$ if $-(\mathbf{A} - t_B\mathbf{C}) \geqslant \mathbf{0}$. Then $t_A < t_\alpha < t_B$ if $0 < \alpha < 1$. From (2.24) and (2.22) it follows immediately that we may take $t_A = q = 0$ and $t_B = q = 4$ in the case of *test(Q)*. In the case of *test(S)*, we have $\mathbf{A}_s - s\mathbf{I}_{(p)} \geqslant \mathbf{0}$ if $s \leqslant \frac{1}{p}$ and $-\mathbf{A}_s + s\mathbf{I}_{(p)} \geqslant \mathbf{0}$ if $s \geqslant 1$, so that we may take $t_A = s = 0$ and $t_B = s = 1$. In the case of *test(V)*, there is no problem with respect to $t_A$, since $\mathbf{A}_\nu - \nu(\mathbf{I}_{(p)} - \mathbf{A}_\nu) \geqslant \mathbf{0}$ if $\nu \leqslant 0$. However, $-\mathbf{A}_\nu + \nu(\mathbf{I}_{(p)} - \mathbf{A}_\nu) \geqslant \mathbf{0}$ is impossible. Here $\mathcal{F}(\nu) = 1$ for $\nu = \infty$. We took $t_A = \nu = 0$ and $t_B = \nu = 1$. Actually, all significance ponts $\nu_\alpha$ relevant to Table 5.8 are such that $0.21 < \nu_{0.05} < 0.24$ and $0.29 < \nu_{0.10} < 0.34$.

## 2.7 Bounds tests

In Durbin and Watson (1950, 1951) a bounds test against autocorrelation in the simple linear model including a constant term regressor has been proposed. In fact, the authors formulated a general lemma which is applicable to all test statistics of the form $T = \hat{\mathbf{u}}'\mathbf{A}\hat{\mathbf{u}}/\hat{\mathbf{u}}'\hat{\mathbf{u}}$ with $\mathbf{A}$ arbitrary. For instance, one may construct a bounds test against hetero variance, or the Durbin-Watson bounds test may be revised so as to include models without a constant term regressor. The essence of a bounds test is its inconclusive region, $[t_l, t_u]$ say. Whereas an exact test using $T$ may have a critical region of the form $T \leqslant t$, so that the acceptance region is $T > t$, the bounds test using the same statistic $T$ has the critical region $T \leqslant t_l$ and the acceptance region $T > t_u$, while the bounds test is inconclusive when the outcome of $T$ falls between $t_l$

and $t_u$.

Below we formulate Lemma 1 for the determination of $t_l$ and $t_u$, by means of *bounding* **X**-*matrices*. It is merely a reformulation of results by Durbin and Watson. For the proof of the lemma, we prefer the approach used in Anderson (1971, section 10.4.2), which is quite different from the approach in Durbin and Watson (1950; see also the corrections to that proof in Durbin and Watson 1951, pp. 177, 178).

*Lemma 1*

Let $\mathbf{X}$ be an arbitrary $n \times k$ matrix and let $T = \hat{\mathbf{u}}'\mathbf{A}\hat{\mathbf{u}}/\hat{\mathbf{u}}'\hat{\mathbf{u}}$ with $\hat{\mathbf{u}} = \mathbf{My} \sim \mathcal{N}(\mathbf{o}, \sigma^2\mathbf{M})$, where $\mathbf{MH}_1 = \mathbf{0}$, $\mathbf{H}_1$ being an $n \times s$ matrix, $(s \leqslant k)$ consisting of $s$ eigenvectors of the symmetric $n \times n$ matrix $\mathbf{A}$ – eventually, $\mathbf{MH}_1 = \mathbf{0}$ is dropped if $s = 0$. From the remaining $n - s$ eigenvectors of $\mathbf{A}$, $n - k$ eigenvectors corresponding to the smallest eigenvalues are the columns of $\mathbf{N}_l$ and $n - k$ eigenvectors corresponding to the greatest eigenvalues are the columns of $\mathbf{N}_u$. Given $\alpha$, $0 < \alpha < 1$, the value of $t$ in $Pr[T \leqslant t] = \alpha$ depends on $\mathbf{X}$ in such a way that, for all $\mathbf{X}$:

$$t_l \leqslant t \leqslant t_u$$

where $t = t_l$ for $\mathbf{M} = \mathbf{N}_l\mathbf{N}_l'$, i.e. $\mathbf{X} = \mathbf{X}_l$ consists of ($k$ linear combinations of) those $k$ eigenvectors of $\mathbf{A}$ which are not included in $\mathbf{N}_l$, and where $t = t_u$ for $\mathbf{M} = \mathbf{N}_u\mathbf{N}_u'$, i.e. $\mathbf{X} = \mathbf{X}_u$ consists of ($k$ linear combinations of) those $k$ eigenvectors of $\mathbf{A}$ which are not included in $\mathbf{N}_u$. For instance, take $n = 10$, $k = 4$, and $\mathbf{A} = \mathbf{A}_v$; see (2.21). We may write $\mathbf{A}_v = \mathbf{I}_{(10)}\mathbf{A}_v\mathbf{I}_{(10)}'$. For $s = 0$, $\mathbf{N}_l$ consists of six eigenvectors corresponding to five zero eigenvalues and one unit eigenvalue, so that $\mathbf{N}_l = [\mathbf{e}_6 \vdots \mathbf{e}_7 \vdots \ldots \vdots \mathbf{e}_{10} \vdots \mathbf{g}]$, where $\mathbf{e}_i$ is the $i$th column of $\mathbf{I}_{(10)}$ and $\mathbf{g}$ is a linear combination of $\mathbf{e}_1, \mathbf{e}_2, \ldots, \mathbf{e}_5$ satisfying $\mathbf{g}'\mathbf{g} = 1$. Taking $\mathbf{g} = \mathbf{e}_5$, we have $\mathbf{X}_l = [\mathbf{e}_1 \vdots \mathbf{e}_2 \vdots \mathbf{e}_3 \vdots \mathbf{e}_4]$. In this example, both $\mathbf{X}_l$ and $\mathbf{X}_u$ are not unique, but all admitted $\mathbf{X}_l$- and $\mathbf{X}_u$-matrices yield the same $t_l$ and $t_u$, respectively. As another instance, take $\mathbf{A} = \mathbf{A}_d$ and consider all $n \times k$ $\mathbf{X}$-matrices such that $\mathbf{Mh}_1^* = \mathbf{o}$; see (2.23), i.e. all $\mathbf{X}$ contain a constant term. Here we find $\mathbf{N}_l = [\mathbf{h}_2^* \vdots \mathbf{h}_3^* \vdots \ldots \vdots \mathbf{h}_{n-k+1}^*]$ and $\mathbf{N}_u = [\mathbf{h}_{k+1}^* \vdots \mathbf{h}_{k+2}^* \vdots \ldots \vdots \mathbf{h}_n^*]$, so that $\mathbf{X}_l = [\mathbf{h}_1^* \vdots \mathbf{h}_{n-k+2}^* \vdots \ldots \vdots \mathbf{h}_n^*]$ and $\mathbf{X}_u = [\mathbf{h}_1^* \vdots \mathbf{h}_2^* \vdots \ldots \vdots \mathbf{h}_k^*]$. Significance points of this test, the Durbin-Watson test, have been tabulated for the latter $\mathbf{X}_l$ and $\mathbf{X}_u$, for several values of $n$, $k$, and $\alpha$ (see Durbin and Watson 1951, or, for more accurate tables, Koerts and Abrahamse 1969). There we may find that the inconclusive region for $n = 15$, $k = 3$, and $\alpha = 0.05$ is [0.946, 1.543]. If we would consider **X**-matrices such that $\mathbf{Mh}_1^* = \mathbf{Mh}_2^* = \mathbf{o}$, then for the same $n$, $k$, and $\alpha$ the inconclusive region would be [1.229, 1.543].

In many apllications with $n \leqslant 20$ the Durbin-Watson bounds test is inconclusive. Durbin and Watson proposed the following procedure: first apply the bounds test and secondly, in inconclusive cases, apply an approximate or exact test. Such a procedure is called a *d*-test. A test is exact if the significance point is calculated exactly, and approximate if the significance point is approximated. Durbin and Watson (1971) compared several approximation methods. For normal situations they recommend the *beta approximation* and the $a + bd_u$ *approximation*, in spite of the fact that more accurate approximations are available which have the disadvantage, however, that they require considerably more computing. Exact significance point calculation requires even more complicated calculations. In Section 5.3 some calculations are carried out on the *d*-test using the beta approximation.

*Proof of Lemma 1*: Let **L** and $\Lambda$ be defined by $\mathbf{N}'\mathbf{A}\mathbf{N} = \mathbf{L}\Lambda\mathbf{L}'$ with $\mathbf{L}' = \mathbf{L}^{-1}$ and $\lambda_1 \leqslant \lambda_2 \leqslant \ldots \leqslant \lambda_{n-k}$, where $\mathbf{N}\mathbf{N}' = \mathbf{M}$ and $\mathbf{N}'\mathbf{N} = \mathbf{I}_{(n-k)}$. Defining $\mathbf{z} = \frac{1}{\sigma}\mathbf{L}'\mathbf{N}'\mathbf{y}$, where $\mathbf{y} \sim \mathcal{n}(\mathbf{X}\boldsymbol{\beta}, \sigma^2\mathbf{I}_{(n)})$, we have $\mathbf{z} \sim \mathcal{n}$ $(\mathbf{o}, \mathbf{I}_{(n-k)})$ and:

$$
\begin{aligned}
T &= \hat{\mathbf{u}}'\mathbf{A}\hat{\mathbf{u}}/\hat{\mathbf{u}}'\hat{\mathbf{u}} \\
&= (\mathbf{N}'\mathbf{y})'\mathbf{N}'\mathbf{A}\mathbf{N}(\mathbf{N}'\mathbf{y}) / (\mathbf{N}'\mathbf{y})'(\mathbf{N}'\mathbf{y}) \\
&= (\mathbf{L}'\mathbf{N}'\mathbf{y})'\Lambda(\mathbf{L}'\mathbf{N}'\mathbf{y}) / (\mathbf{L}'\mathbf{N}'\mathbf{y})'(\mathbf{L}'\mathbf{N}'\mathbf{y}) \\
&= \mathbf{z}'\Lambda\mathbf{z}/\mathbf{z}'\mathbf{z}
\end{aligned}
$$

i.e. $T$ is the ratio of a weighted and the nonweighted sum of $n$-$k$ mutually independent $\chi^2(1)$-variables, the weights in the numerator being the eigenvalues of $\mathbf{N}'\mathbf{A}\mathbf{N}$. Now writing $\mathbf{A} = \mathbf{H}\mathbf{D}\mathbf{H}'$ with $\mathbf{H}' = \mathbf{H}^{-1}$ and $\mathbf{H} = [\mathbf{H}_1 \vdots \mathbf{H}_2]$, then, for all **N** such that $\mathbf{N}'\mathbf{H}_1 = \mathbf{0}$, we have $\mathbf{N}'\mathbf{A}\mathbf{N} = \mathbf{N}'\mathbf{H}\mathbf{D}\mathbf{H}'\mathbf{N} = (\mathbf{H}_2'\mathbf{N})'\mathbf{D}^*(\mathbf{H}_2'\mathbf{N})$, where $(\mathbf{H}_2'\mathbf{N})'(\mathbf{H}_2'\mathbf{N}) = \mathbf{I}_{(n-k)}$ in view of $\mathbf{I}_{(n-k)} = \mathbf{N}'\mathbf{N} = \mathbf{N}'\mathbf{H}\mathbf{H}'\mathbf{N} = [\mathbf{0} \vdots \mathbf{N}'\mathbf{H}_2][\mathbf{0} \vdots \mathbf{N}'\mathbf{H}_2]' = (\mathbf{H}_2'\mathbf{N})'(\mathbf{H}_2'\mathbf{N})$, and where $\mathbf{D}^*$ contains the eigenvalues of **A** corresponding to the eigenvectors in $\mathbf{H}_2$. We assume that $d_1^* \leqslant d_2^* \leqslant \ldots \leqslant d_{n-s}^*$. Lemma 2 below states that:

$$d_i^* \leqslant \lambda_i \leqslant d_{k-s+i}^* \qquad i = 1, 2, \ldots, n\text{-}k$$

Let $\Lambda_l$, $\Lambda$, and $\Lambda_u$ be diagonal matrices of order $(n\text{-}k) \times (n\text{-}k)$ and with $i$th diagonal element $d_i^*$, $\lambda_i$, and $d_{k-s+i}^*$, respectively, then we have $\Lambda_l \leqslant \Lambda \leqslant \Lambda_u$. It is easily verified that $\Lambda_l = \mathbf{N}_l'\mathbf{A}\mathbf{N}_l$ and $\Lambda_u = \mathbf{N}_u'\mathbf{A}\mathbf{N}_u$. Hence, for arbitrary $(n\text{-}k)$-element vectors $\mathbf{z}$ we have $\mathbf{z}'\Lambda_l\mathbf{z} \leqslant \mathbf{z}'\Lambda\mathbf{z} \leqslant \mathbf{z}'\Lambda_u\mathbf{z}$, and also $T_l \leqslant T \leqslant T_u$, where $T_l = \mathbf{z}'\Lambda_l\mathbf{z}/\mathbf{z}'\mathbf{z}$ and $T_u = \mathbf{z}'\Lambda_u\mathbf{z}/\mathbf{z}'\mathbf{z}$. In particular, when $\mathbf{z}$ is a vector of random variables, whose distribution does not depend on $\mathbf{X}$, then $Pr[T_l \leqslant t_l] = Pr[T \leqslant t] = Pr[T_u \leqslant t_u] = \alpha$ implies $t_l \leqslant t \leqslant t_u$ ($T_l$ and $T_u$ are called the lower and the upper bound statistic).

*Lemma 2*

Let $d_1^* \leqslant d_2^* \leqslant \ldots \leqslant d_{n-s}^*$ be the diagonal elements of $\mathbf{D}^*$; let $\mathbf{G}$ be an $(n\text{-}s) \times (n\text{-}k)$ matrix satisfying $\mathbf{G}'\mathbf{G} = \mathbf{I}_{(n-k)}$; let $\lambda_1 \leqslant \lambda_2 \leqslant \ldots \leqslant \lambda_{n-k}$ be the eigenvalues of $\mathbf{G}'\mathbf{D}^*\mathbf{G}$; then:

$$d_i^* \leqslant \lambda_i \leqslant d_{k-s+i}^* \qquad i = 1, 2, \ldots, n\text{-}k$$

*Proof of Lemma 2*: Let $\mathbf{q}$ be an arbitrary nonzero vector satisfying $\mathbf{q}'\mathbf{e}_j = 0$ for $j = 1, 2, \ldots, i\text{-}1$, where $\mathbf{e}_j$ is the $j$th column of $\mathbf{I}_{(n-s)}$. Hence, $\mathbf{q}$ is perpendicular to the $(i - 1)$-dimensional space $\mathcal{M}(\mathbf{e}_1 \vdots \mathbf{e}_2 \vdots \ldots \vdots \mathbf{e}_{i-1})$ and $\mathbf{q}$ is a linear combination of $\mathbf{e}_i, \mathbf{e}_{i+1}, \ldots, \mathbf{e}_{n-s}$, say $\mathbf{q} = \sum_{j=1}^{n-s} \alpha_j \mathbf{e}_j$. Then $min(\mathbf{q}'\mathbf{D}^*\mathbf{q}/\mathbf{q}'\mathbf{q}) = min(\sum_{j=i}^{n-s} \alpha_j^2 d_j^* / \sum_{j=i}^{n-s} \alpha_j^2) = d_i^*$. Further, let $\mathbf{q} = \mathbf{G}\mathbf{G}'\mathbf{q}$, so that $\mathbf{q}$ is perpendicular to the space $\mathcal{M}(\mathbf{I} - \mathbf{G}\mathbf{G}' \vdots \mathbf{e}_1 \vdots \mathbf{e}_2 \vdots \ldots \vdots \mathbf{e}_{i-1})$, where $\mathbf{I} - \mathbf{G}\mathbf{G}'$ has rank $k\text{-}s$. Since $\mathbf{G}$ in the lemma is arbitrary, the dimension of the latter space may be equal to $k\text{-}s\text{+}i\text{-}1$. Hence, $k\text{-}s\text{+}i\text{-}1 < n\text{-}s$, otherwise $\mathbf{q} = \mathbf{o}$. It follows that $i \leqslant n\text{-}k$. Now $\mathbf{q}'\mathbf{e}_j = \mathbf{q}'\mathbf{G}\mathbf{G}'\mathbf{e}_j = \mathbf{v}'\mathbf{c}_j$, where $\mathbf{v} = \mathbf{G}'\mathbf{q}$ and $\mathbf{c}_j = \mathbf{G}'\mathbf{e}_j$, and we have:

$$d_i^* = \min_{\substack{\mathbf{q}'\mathbf{e}_j=0 \\ j=1,\ldots,i-1}} \left(\frac{\mathbf{q}'\mathbf{D}^*\mathbf{q}}{\mathbf{q}'\mathbf{q}}\right) \leqslant \min_{\substack{\mathbf{q}'\mathbf{e}_j=0 \\ j=1,\ldots,i-1 \\ \mathbf{q}=\mathbf{G}\mathbf{G}'\mathbf{q}}} \left(\frac{\mathbf{q}'\mathbf{D}^*\mathbf{q}}{\mathbf{q}'\mathbf{q}}\right) =$$

$$\min_{\substack{\mathbf{q}'\mathbf{e}_j=0 \\ j=1,\ldots,i-1}} \left(\frac{\mathbf{q}'\mathbf{G}\mathbf{G}'\mathbf{D}^*\mathbf{G}\mathbf{G}'\mathbf{q}}{\mathbf{q}'\mathbf{G}\mathbf{G}'\mathbf{q}}\right) = \min_{\substack{\mathbf{v}'\mathbf{c}_j=0 \\ j=1,\ldots,i-1}} \left(\frac{\mathbf{v}'\mathbf{G}'\mathbf{D}^*\mathbf{G}\mathbf{v}}{\mathbf{v}'\mathbf{v}}\right) \leqslant \lambda_i$$

The final inequality follows from Lemma 3 below. Analogously:

$$d^*_{k-s+i} = \max_{\substack{q'e_j=0 \\ j=k-s+i+1,\dots,n-s}} \left(\frac{q'D^*q}{q'q}\right) \geqslant \max_{\substack{q'e_j=0 \\ j=k-s+i+1,\dots,n-s \\ q=GG'q}} \left(\frac{q'D^*q}{q'q}\right) =$$

$$\max_{\substack{q'e_j=0 \\ j=k-s+i+1,\dots,n-s}} \left(\frac{q'GG'D^*GG'q}{q'GG'q}\right) = \max_{\substack{v'c_j=0 \\ j=k-s+i+1,\dots,n-s}} \left(\frac{v'G'D^*Gv}{v'v}\right) \geqslant \lambda_i$$

Again the final inequality follows from Lemma 3, after renumbering $c_j$ as $c_h$ with $h = j-k+s$, while $p = n-k$. In order that q be nonzero we must have $i > 0$, so that, for the two parts together, we have $1 \leqslant i \leqslant n-k$.

*Lemma 3*
Let **A** be a symmetric $p \times p$ matrix with eigenvalues $\lambda_1 \leqslant \lambda_2 \leqslant \dots \leqslant \lambda_p$ and let $c_1, \dots, c_{i-1}, c_{i+1}, \dots, c_p$ be arbitrarily fixed $p$-element vectors, then:

$$\min_{\substack{v'c_j=0 \\ j=1,\dots,i-1}} \left(\frac{v'Av}{v'v}\right) \leqslant \lambda_i \leqslant \max_{\substack{v'c_j=0 \\ j=i+1,\dots,p}} \left(\frac{v'Av}{v'v}\right)$$

*Proof of Lemma 3*: Writing $A = L\Lambda L'$ with $L' = L^{-1}$, and defining $z = L'v$ and $b_j = L'c_j$, we have:

$$\min_{\substack{v'c_j=0 \\ j=1,\dots,i-1}} \left(\frac{v'L\Lambda L'v}{v'LL'v}\right) = \min_{\substack{z'b_j=0 \\ j=1,\dots,i-1}} \left(\frac{z'\Lambda z}{z'z}\right) =$$

$$\leqslant \min_{\substack{z'b_j=0 \\ j=1,\dots,i-1 \\ z_{i+1}=\dots=z_p=0}} \left(\frac{z'\Lambda z}{z'z}\right) = \min_{\substack{z'b_j=0 \\ j=1,\dots,i-1}} \left(\frac{\sum_{t=1}^{i} \lambda_t z_t^2}{\sum_{t=1}^{i} z_t^2}\right) \leqslant \lambda_i$$

Note that $z'b_j = 0$ for $j = 1, 2, \dots, i-1$ implies that z be perpendicular to the space $\mathfrak{M}(b_1 : b_2 :\dots: b_{i-1})$, which is arbitrary and may have dimension $i-1$. If $z_{h+1} = z_{h+2} = \dots = z_p = 0$, then z is perpendicular to the fixed $(p-h)$-dimensional space $\mathfrak{M}(e_{h+1} : e_{h+2} :\dots: e_p)$. For some set of vectors $b_1, b_2, \dots, b_{i-1}$ the two restrictions together imply that z be perpendicular to a $(p-h+i-1)$-dimensional space. Hence

$p$-$h$+$i$-1 $\leqslant p$-1, otherwise $\mathbf{z} = \mathbf{o}$; i.e. $h \geqslant i$. Therefore, the above additional restriction $z_{i+1} = \ldots = z_p = 0$ cannot be extended to more elements of $\mathbf{z}$. Analogously:

$$\max_{\substack{\mathbf{v}'\mathbf{c}_j=0 \\ j=i+1,\ldots,p}} \left(\frac{\mathbf{v}'\mathbf{L}\Lambda\mathbf{L}'\mathbf{v}}{\mathbf{v}'\mathbf{L}\mathbf{L}'\mathbf{v}}\right) = \max_{\substack{\mathbf{z}'\mathbf{b}_j=0 \\ j=i+1,\ldots,p}} \left(\frac{\mathbf{z}'\Lambda\mathbf{z}}{\mathbf{z}'\mathbf{z}}\right) =$$

$$\geqslant \max_{\substack{\mathbf{z}'\mathbf{b}_j=0 \\ j=i+1,\ldots,p \\ z_1=\ldots=z_{i-1}=0}} \left(\frac{\mathbf{z}'\Lambda\mathbf{z}}{\mathbf{z}'\mathbf{z}}\right) = \max_{\substack{\mathbf{z}'\mathbf{b}_j=0 \\ j=i+1,\ldots,p}} \left(\frac{\sum_{t=i}^{p}\lambda_t z_t^2}{\sum_{t=i}^{p} z_t^2}\right) \geqslant \lambda_i$$

# 3. *BLUF* disturbance estimation

## 3.1 The problem

For the purpose of testing hypotheses concerning parameters of $\Gamma$, we wish to have a vector $\mathbf{w}$ satisfying (2.14). There is a whole class of such vectors, as we shall see. We are, of course, interested in the vector which gives best testing results, i.e. maximal power. Unfortunately, we did not succeed in translating the maximal power criterion into a manageable optimality criterion for $\mathbf{w}$. Therefore we adopt a criterion based on other considerations.

In Section 2.4, we proposed to choose $\mathbf{w}$ as close as possible to $\mathbf{J}'\mathbf{u}$, where $\mathbf{J}$ is an $n \times p$ matrix independent of $\mathbf{y}$. At this moment we do not specify $\mathbf{J}$, although we have several specifications in mind, namely $\mathbf{J} = \mathbf{I}_{(n)}$, in which case $\mathbf{w}$ is as close as possible to $\mathbf{u}$, or $\mathbf{J} = \mathbf{M}^{*\prime}$, in which case $\mathbf{w}$ is as close as possible to $\mathbf{u}^*$, or $\mathbf{J}$ is an $n \times (n\text{-}k)$ submatrix of $\mathbf{I}_{(n)}$, in order to include the theory of the *BLUS* vectors (see Section 3.3). Analogous to $\mathbf{u}^*$ in Section 1.3, the vector $\mathbf{w}$ is that vector $\mathbf{v}$ which minimizes:

$$\mathcal{E}[(\mathbf{v} - \mathbf{J}'\mathbf{u})'\mathbf{Q}(\mathbf{v} - \mathbf{J}'\mathbf{u})] \tag{3.1}$$

subject to $\mathbf{v} = \mathbf{B}'\mathbf{y}$ with $\mathbf{B}$ independent of $\mathbf{y}$, $\mathbf{B}'\mathbf{X} = \mathbf{0}$, and $\mathbf{B}'\Gamma\mathbf{B} = \Omega$. Here $\mathbf{Q}$ is an arbitrary matrix satisfying $\mathbf{Q} = \mathbf{Q}'_{(p)} > \mathbf{0}$, and $\Omega = \mathbf{K}\mathbf{K}'$, where $\mathbf{K}$ is a fixed arbitrary $p \times r$ matrix with rank $r$. The specification of $\mathbf{K}$ is the subject of Chapter 4. We call the $p$-element vector $\mathbf{w}$ the *BLUF* (*F*: *fixed* covariance matrix) disturbance estimator, for the same reason as $\mathbf{u}^*$ is called the *BLU* disturbance estimator (see Section 1.3). Given $\mathcal{E}(\mathbf{y}) = \mathbf{X}\boldsymbol{\beta}$ and $\mathcal{V}(\mathbf{y}) = \sigma^2\Gamma$, it is easily seen that $\mathcal{E}(\mathbf{v}) = \mathbf{o}$ and $\mathcal{V}(\mathbf{v}) = \sigma^2\Omega$. Thus, when we say that $\mathbf{v}$ has a fixed covariance matrix, we mean that this matrix is fixed apart from scalar multiplication, since $\sigma^2$ is unknown. The only restriction on the positive integers $p$ and $r$ is $r \leqslant min\ (p, n\text{-}k)$, see (2.14). In the next section we derive that:

$$\mathbf{w} = \mathbf{K}(\mathbf{K}'\mathbf{QJ}'\mathbf{M}^*\mathbf{\Gamma JQK})^{-\frac{1}{2}}\mathbf{K}'\mathbf{QJ}'\mathbf{M}^*\mathbf{y} \qquad (3.2)$$

This vector uniquely minimizes (3.1), provided that $\mathbf{K}'\mathbf{QJ}'\mathbf{M}^*\mathbf{\Gamma JQK}$ is nonsingular. The case, that the latter matrix product is singular, is considered in the next section. Note that **w** is a linear transformation of $\mathbf{u}^* = \mathbf{M}^*\mathbf{y}$. This is not surprising, since we required that $\mathbf{v} = \mathbf{B}'\mathbf{y}$ with $\mathbf{B}'\mathbf{X} = \mathbf{0}$, so that the columns of **B** are vectors lying in $\mathcal{M}(\mathbf{X})^{\perp} = \mathcal{M}(\mathbf{M}) = \mathcal{M}(\mathbf{M}^{*\prime})$, i.e. $\mathbf{B} = \mathbf{M}^{*\prime}\mathbf{G}$ for some matrix **G**. Further note that the *BLUF* estimator of $\mathbf{u}^*$ (i.e. $\mathbf{J}' = \mathbf{M}^*$) and the *BLUF* estimator of **u** (i.e. $\mathbf{J} = \mathbf{I}$) coincide, since $\mathbf{J}'\mathbf{M}^* = \mathbf{M}^*\mathbf{M}^* = \mathbf{M}^*$ and $\mathbf{M}^*\mathbf{\Gamma J} = \mathbf{M}^*\mathbf{\Gamma M}^{*\prime} = \mathbf{M}^*\mathbf{\Gamma}$ when $\mathbf{J}' = \mathbf{M}^*$.

## 3.2 The derivation of w

We derive the *BLUF* estimator **w** of $\mathbf{J}'\mathbf{u}$ in two steps: we first find the class of *LUF* vectors, and from this class we select the best estimator, in accordance with (3.1).

*L*

The general form of vectors **v**, which are linear functions of **y**, is $\mathbf{v} = \mathbf{B}'\mathbf{y} + \mathbf{c}$, where **B** and **c** are independent of **y**. Strict linearity requires $\mathbf{c} = \mathbf{o}$ (see the remark in Section 1.2).

*LU*

$\mathbf{B}'\mathbf{y} + \mathbf{c}$ has zero expectation for all values of the unknown $\boldsymbol{\beta}$ if and only if $\mathbf{B}'\mathbf{X} = \mathbf{0}$ and $\mathbf{c} = \mathbf{o}$, in view of $\mathcal{E}(\mathbf{B}'\mathbf{y} + \mathbf{c}) = \mathbf{B}'\mathbf{X}\boldsymbol{\beta} + \mathbf{c}$. It follows that $\mathbf{v} = \mathbf{B}'\mathbf{y} = \mathbf{B}'\mathbf{u}$ and that all $p$ columns of **B** are vectors lying in $\mathcal{M}(\mathbf{X})^{\perp}$. The rows of $\overline{\mathbf{N}}'\mathbf{S}^{-1}$ span $\mathcal{M}(\mathbf{X})^{\perp}$, since $\overline{\mathbf{N}}'\overline{\mathbf{X}} = \overline{\mathbf{N}}'\mathbf{S}^{-1}\mathbf{X} = \mathbf{0}$ and $rank(\overline{\mathbf{N}}'\mathbf{S}^{-1}) = n\text{-}k$ (see Section 1.3). Thus $\mathbf{B}'\mathbf{X} = \mathbf{0}$ if and only if $\mathbf{B}' = \mathbf{F}\overline{\mathbf{N}}'\mathbf{S}^{-1}$ for some $p \times (n\text{-}k)$ matrix **F**.

*LUF*

$\mathbf{B}'\mathbf{y}$ has covariance matrix $\sigma^2\mathbf{\Omega} = \sigma^2\mathbf{KK}'$ if and only if $\mathbf{B}'\mathbf{SS}'\mathbf{B} = \mathbf{KK}'$. In view of $\mathbf{B}' = \mathbf{F}\overline{\mathbf{N}}'\mathbf{S}^{-1}$ we have $\mathbf{KK}' = \mathbf{B}'\mathbf{SS}'\mathbf{B} = \mathbf{FF}'$. In accordance with Result 1.A.6.3, $\mathbf{F} = \mathbf{KH}'$ for some orthogonal $(n\text{-}k) \times r$

matrix **H**.

Summarizing the first step, all vectors $\mathbf{KH}'\bar{\mathbf{N}}'\mathbf{S}^{-1}\mathbf{y}$ with **H** orthogonal together form the entire class of *LUF* vectors with covariance matrix $\sigma^2\Omega$. From this class we choose the one which minimizes (3.1). This must evidently be achieved by an appropriate choice of **H**.

*BLUF*

We minimize:

$$\begin{aligned}\mathcal{E}[(\mathbf{v}-\mathbf{J}'\mathbf{u})'\mathbf{Q}(\mathbf{v}-\mathbf{J}'\mathbf{u})] &= \mathcal{E}[\mathbf{u}'(\mathbf{B}-\mathbf{J})\mathbf{Q}(\mathbf{B}-\mathbf{J})'\mathbf{u}]\\ &= \sigma^2 tr[(\mathbf{B}-\mathbf{J})\mathbf{Q}(\mathbf{B}-\mathbf{J})'\Gamma]\\ &= \sigma^2 tr(\mathbf{QB}'\Gamma\mathbf{B}) + \sigma^2 tr(\mathbf{QJ}'\Gamma\mathbf{J}) -\\ &\quad 2\sigma^2 tr(\mathbf{B}'\Gamma\mathbf{JQ})\\ &= \sigma^2 tr(\mathbf{Q}\Omega) + \sigma^2 tr(\mathbf{QJ}'\Gamma\mathbf{J}) -\\ &\quad 2\sigma^2 tr(\mathbf{H}'\bar{\mathbf{N}}'\mathbf{S}'\mathbf{JQK})\end{aligned}$$

with respect to **H**. Hence, the problem is to maximize $tr(\mathbf{H}'\mathbf{W})$ with respect to **H**, subject to $\mathbf{H}'\mathbf{H} = \mathbf{I}$, where $\mathbf{W} = \bar{\mathbf{N}}'\mathbf{S}'\mathbf{JQK}$. In accordance with Result 1.A.6.8, the solution matrix **H** is $\mathbf{H}_0$:

$$\mathbf{H}_0 = \mathbf{UT}$$

where **T** is an orthogonal $r \times r$ matrix whose columns are eigenvectors of $\mathbf{W}'\mathbf{W}$, and **U** is an orthogonal $(n\text{-}k) \times r$ matrix whose columns are eigenvectors of $\mathbf{WW}'$, such that $\mathbf{W} = \mathbf{U}\Lambda^{1/2}\mathbf{T}'$, where $\Lambda$ is the matrix of eigenvalues of $\mathbf{W}'\mathbf{W}$. The construction of **U** and **T** follows from the proof of Result 1.A.6.7.

The formula $\mathbf{H}_0 = \mathbf{UT}'$ is general. In particular, it holds for both $rank(\mathbf{W}) < r$ and $rank(\mathbf{W}) = r$, •—• $rank(\mathbf{W}) > r$ being impossible. The case, where $rank(\mathbf{W}) < r$, turns out to be of very little practical interest. We shall not meet this case in our applications. In the other case, where $rank(\mathbf{W}) = r$, we have $rank(\Lambda) = r$, so that $\Lambda^{-1}$ exists, and hence:

$$\mathbf{H}_0 = \mathbf{UT}' = \mathbf{U}\Lambda^{1/2}\mathbf{T}'\mathbf{T}\Lambda^{-1/2}\mathbf{T}' = \mathbf{W}(\mathbf{W}'\mathbf{W})^{-1/2}$$

Now we have:

$$\mathbf{w} = \mathbf{B}'\mathbf{y} = \mathbf{K}\mathbf{H}_0'\bar{\mathbf{N}}'\mathbf{S}^{-1}\mathbf{y} = \mathbf{K}(\mathbf{W}'\mathbf{W})^{-\frac{1}{2}}\mathbf{W}'\bar{\mathbf{N}}'\mathbf{S}^{-1}\mathbf{y}$$

$$= \mathbf{K}(\mathbf{K}'\mathbf{Q}\mathbf{J}'\mathbf{S}\bar{\mathbf{N}}\bar{\mathbf{N}}'\mathbf{S}'\mathbf{J}\mathbf{Q}\mathbf{K})^{-\frac{1}{2}}\mathbf{K}'\mathbf{Q}\mathbf{J}'\mathbf{S}\bar{\mathbf{N}}\bar{\mathbf{N}}'\mathbf{S}^{-1}\mathbf{y}$$

$$= \mathbf{K}(\mathbf{K}'\mathbf{Q}\mathbf{J}'\mathbf{M}^*\Gamma\mathbf{J}\mathbf{Q}\mathbf{K})^{-\frac{1}{2}}\mathbf{K}'\mathbf{Q}\mathbf{J}'\mathbf{M}^*\mathbf{y}$$

From Result 1.A.6.8 we know that $\mathbf{H}_0$ uniquely maximizes $tr(\mathbf{H}'\mathbf{W})$ subject to $\mathbf{H}'\mathbf{H} = \mathbf{I}$, when $\mathbf{W}$ has rank $r$. However, $\mathbf{B}' = \mathbf{K}\mathbf{H}_0'\bar{\mathbf{N}}'\mathbf{S}^{-1}$, where $\mathbf{K}$, $\bar{\mathbf{N}}$, and $\mathbf{S}$ are not unique for given $\Omega$, $\mathbf{X}$ and $\Gamma$. That is, orthogonal transformations of the columns of $\mathbf{K}$, $\bar{\mathbf{N}}$, and $\mathbf{S}$ are admissible. It turns out that $\mathbf{w}$ is invariant under such transformations, which can be seen as follows.

Observe that $\bar{\mathbf{N}}$ and $\mathbf{S}$ in $\mathbf{w}$ occur only as $\mathbf{S}\bar{\mathbf{N}}\bar{\mathbf{N}}'\mathbf{S}'$ and $\mathbf{S}\bar{\mathbf{N}}\bar{\mathbf{N}}'\mathbf{S}^{-1} = \mathbf{S}\bar{\mathbf{N}}\bar{\mathbf{N}}'\mathbf{S}'\Gamma^{-1}$. Clearly $\mathbf{w}$ is invariant under replacement of $\bar{\mathbf{N}}$ by $\bar{\mathbf{N}}\mathbf{T}_n'$ with $\mathbf{T}_n$ satisfying $\mathbf{T}_n' = \mathbf{T}_n^{-1}$. We have $\mathbf{S}\bar{\mathbf{N}}\bar{\mathbf{N}}'\mathbf{S}' = \mathbf{S}\bar{\mathbf{M}}\mathbf{S}' = \mathbf{M}^*\Gamma$ (see Section 1.3), from which it is clear that replacement of $\mathbf{S}$ by $\mathbf{S}\mathbf{T}_s'$ with $\mathbf{T}_s$ satisfying $\mathbf{T}_s' = \mathbf{T}_s^{-1}$ is irrelevant for $\mathbf{w}$. It remains to be shown that $\mathbf{w}$ is invariant under replacement of $\mathbf{K}$ by $\mathbf{K}\mathbf{T}_k'$ with $\mathbf{T}_k$ satisfying $\mathbf{T}_k' = \mathbf{T}_k^{-1}$. Such a replacement in $\mathbf{K}(\mathbf{K}' \ldots \mathbf{K})^{-\frac{1}{2}}\mathbf{K}' \ldots$ yields $\mathbf{K}\mathbf{T}_k'(\mathbf{T}_k\mathbf{K}' \ldots \mathbf{K}\mathbf{T}_k')^{-\frac{1}{2}}\mathbf{T}_k\mathbf{K}' \ldots$, which expression is equal to $\mathbf{K}\mathbf{T}_k'\mathbf{T}_k(\mathbf{K}' \ldots \mathbf{K})^{-\frac{1}{2}}\mathbf{T}_k'\mathbf{T}_k\mathbf{K}' \ldots$ (see Result 1.A.6.6), and $\mathbf{T}_k'\mathbf{T}_k = \mathbf{I}$. Consequently, when $\mathbf{X}$, $\Gamma$, $\Omega$, $\mathbf{J}$, and $\mathbf{Q}$ are given, the *BLUF* estimator $\mathbf{w}$ in (3.2) is unique.

*Rank*($\mathbf{W}$) = $r$ cannot be ensured by individual rank requirements for $\mathbf{X}$, $\Gamma$, $\Omega$, $\mathbf{J}$, and $\mathbf{Q}$. For instance, if $\Gamma = \mathbf{Q} = \mathbf{J} = \mathbf{I}_{(n)}$ and the $n \times (n\text{-}k)$ matrix $\mathbf{K}$ has rank $n\text{-}k$ (we always consider $n \times k$ $\mathbf{X}$-matrices with rank $k$), then all individual ranks are maximal, and $\mathbf{W}'\mathbf{W} = \mathbf{K}'\mathbf{M}\mathbf{K}$. The rank of $\mathbf{W}'\mathbf{W}$, and hence the rank of $\mathbf{W}$, falls below $r(= n\text{-}k)$ if $\mathbf{K}$ and $\mathbf{X}$ have one or more columns in common, in view of $\mathbf{M}\mathbf{X} = \mathbf{0}$. In this case one or more eigenvalues of $\mathbf{K}'\mathbf{M}\mathbf{K}$ are equal to zero. However, in Chapter 4 we adopt a criterion for the specification of $\mathbf{K}$ which is based on maximization of the (nonnegative) eigenvalues of $\mathbf{K}'\mathbf{M}\mathbf{K}$. One or more zero eigenvalues is practically impossible.

We have already observed the fact that $\mathbf{w}$ is the same for $\mathbf{J} = \mathbf{I}_{(n)}$ with *rank*($\mathbf{J}$) = $n$ and for $\mathbf{J}' = \mathbf{M}^*$ with *rank*($\mathbf{J}$) = $n - k$. It is easily verified that $\mathbf{w}$ is also the same when $\mathbf{Q} = \mathbf{I}_{(n)}$ with *rank* ($\mathbf{Q}$) = $n$ is replaced by $\mathbf{Q} = \Omega$ with *rank* ($\mathbf{Q}$) = $r \leqslant n\text{-}k$. From this paragraph and the preceding one it follows that maximal rank of $\mathbf{J}$ and $\mathbf{Q}$ is neither a sufficient condition nor a necessary condition for the existence of $\mathbf{w}$ as in (3.2). Of course, neither *rank*(**J**) nor *rank*(**Q**) should be smaller than *rank*(**K**) = r.

## 3.3 Special cases

For particular specifications of Γ, Ω, **J**, and **Q** in (3.2) some familiar estimators appear:

*The BLU estimators.*
When $\mathbf{J} = \mathbf{I}_{(n)}$, (3.2) becomes:

$$\mathbf{w} = \mathbf{K}(\mathbf{K}'\mathbf{Q}\mathbf{S}\bar{\mathbf{N}}\bar{\mathbf{N}}'\mathbf{S}'\mathbf{Q}\mathbf{K})^{-\frac{1}{2}}\,\mathbf{K}'\mathbf{Q}\mathbf{S}\bar{\mathbf{N}}\bar{\mathbf{N}}'\mathbf{S}^{-1}\mathbf{y}$$

Taking $\mathbf{K} = \mathbf{S}\bar{\mathbf{N}}$, so that $\Omega = \mathbf{S}\bar{\mathbf{N}}\bar{\mathbf{N}}'\mathbf{S}' = \mathbf{M}^*\Gamma$ (see Section 1.3), we find:

$$\begin{aligned}\mathbf{w} &= \mathbf{S}\bar{\mathbf{N}}(\bar{\mathbf{N}}'\mathbf{S}'\mathbf{Q}\mathbf{S}\bar{\mathbf{N}})^{-1}\bar{\mathbf{N}}'\mathbf{S}'\mathbf{Q}\mathbf{S}\bar{\mathbf{N}}\bar{\mathbf{N}}'\mathbf{S}^{-1}\mathbf{y} \\ &= \mathbf{S}\bar{\mathbf{N}}\bar{\mathbf{N}}'\mathbf{S}^{-1}\mathbf{y} = \mathbf{M}^*\mathbf{y} = \mathbf{u}^*\end{aligned}$$

If besides Γ = **I**, then we may take **S** = **I**, so that $\bar{\mathbf{N}} = \mathbf{N}$, $\mathbf{K} = \mathbf{N}$, $\Omega = \mathbf{M}$, and:

$$\mathbf{w} = \mathbf{M}\mathbf{y} = \hat{\mathbf{u}}$$

*The BLUS estimators.*
When $\Gamma = \mathbf{I}_{(n)}$, $\mathbf{Q} = \Omega = \mathbf{I}_{(n-k)}$, and **J** is an $n \times (n\text{-}k)$ matrix obtained from $\mathbf{I}_{(n)}$ by deleting $k$ columns, then:

$$\mathbf{w} = (\mathbf{J}'\mathbf{M}\mathbf{J})^{-\frac{1}{2}}\,\mathbf{J}'\mathbf{M}\mathbf{y}$$

which vector is called the *BLUS*(*S* : *scalar* fixed covariance matrix) estimator of $\mathbf{J}'\mathbf{u}$ (see Theil 1971).

*The new estimators.*
Disappointed by the considerable loss of power in autocorrelation tests using *BLUS* estimators, compared with Durbin-Watson tests, Abrahamse and Koerts (1971) developed the so-called new estimators. Attributing the loss of power to the difference between $\sigma^2\mathbf{M}$ and

$\sigma^2 \mathbf{I}_{(n\text{-}k)}$, the covariance matrices of $\hat{\mathbf{u}}$ and the *BLUS* estimators, respectively, they formulated the *BLUF* estimation problem with $\Gamma = \mathbf{J} = \mathbf{Q} = \mathbf{I}_{(n)}$ and $\Omega = \Omega^2_{(n)}$. They found:

$$\mathbf{w} = \mathbf{K}(\mathbf{K}'\mathbf{M}\mathbf{K})^{-1/2}\mathbf{K}'\mathbf{M}\mathbf{y}$$

with $\mathbf{K}$ satisfying $\mathbf{K}'\mathbf{K} = \mathbf{I}_{(n\text{-}k)}$. This vector seems very suitable for practical applications. We make frequent use of it in Chapter 5. For matrices $\mathbf{K}$ consisting of $n$-$k$ distinct columns of $\mathbf{I}_{(n)}$, like the matrix $\mathbf{J}$ in the *BLUS* vector above, we call $\mathbf{w}$ a *modified BLUS* estimator. For instance, if the first $k$ rows of such a matrix $\mathbf{K}$ are all zero, then the first $k$ elements of $\mathbf{w}$ are zeros and the last $(n$-$k)$-element subvector of $\mathbf{w}$ is equal to the *BLUS* estimator of $\mathbf{K}'\mathbf{u}$.

In Section 3.5 we consider another disturbance estimator which is *LUF*. That vector is not best in the sense of (3.1), but it has the advantage that it is much more easy to calculate. Powers of tests based on this vector are also compared in Chapter 5.

## 3.4 The residual aspect

We examine under what conditions a *LU* disturbance estimator can be regarded as a regression residual vector, and we consider the implications for $\mathbf{w}$.

In the linear model (1.1) we call $\mathbf{v}$ a regression residual vector if and only if $\mathbf{v} = \mathbf{y} - \mathbf{X}\mathbf{b}$ holds for all $\mathbf{y}$, where the $k$-element vector $\mathbf{b}$ is a function of $\mathbf{y}$. Writing any *LU* disturbance estimator as $\mathbf{B}'\mathbf{y}$ with $\mathbf{B}'\mathbf{X} = \mathbf{0}$, then $\mathbf{B}'\mathbf{y}$ is a regression residual vector if and only if $\mathbf{y} = \mathbf{X}\mathbf{b} + \mathbf{B}'\mathbf{y}$ with $\mathbf{B}'\mathbf{X} = \mathbf{0}$ holds for all $\mathbf{y}$. It follows from $\mathbf{B}'\mathbf{X} = \mathbf{0}$ that $rank(\mathbf{B}) \leqslant n$-$k$, while, in view of $\mathbf{y} = \mathbf{X}\mathbf{b} + \mathbf{B}'\mathbf{y}$, the columns of $\mathbf{X}$ and $\mathbf{B}'$ must span the whole $n$-dimensional space, implying $rank(\mathbf{B}) \geqslant n - k$. Hence, $rank(\mathbf{B}) = n$-$k$, and $\mathbf{B}$ is a square matrix. Premultiplication of $\mathbf{y} = \mathbf{X}\mathbf{b} + \mathbf{B}'\mathbf{y}$ by $\mathbf{B}'$ yields $\mathbf{B}'\mathbf{y} = \mathbf{o} + \mathbf{B}'\mathbf{B}'\mathbf{y}$, i.e. $\mathbf{B} = \mathbf{B}^2$. Hence, $\mathbf{B}'\mathbf{y}$ with $\mathbf{B}'\mathbf{X} = \mathbf{0}$ is a regression residual vector only if $\mathbf{B}$ is an idempotent $n \times n$ matrix with rank $n$-$k$. Conversely, if $\mathbf{B}$ is an idempotent $n \times n$ matrix with rank $n$-$k$, satisfying $\mathbf{B}'\mathbf{X} = \mathbf{0}$, then $\mathfrak{M}(\mathbf{X}) = \mathfrak{M}(\mathbf{B})^{\perp}$, in view of the ranks, and $\mathfrak{M}(\mathbf{X}) = \mathfrak{M}(\mathbf{I} - \mathbf{B}')$ – see Result 1.A.6.2. Hence, for every vector $\mathbf{y}$ there exists a unique $k$-element vector $\mathbf{b}$ such that $(\mathbf{I} - \mathbf{B}')\mathbf{y} = \mathbf{X}\mathbf{b}$, so that $\mathbf{B}'\mathbf{y} = \mathbf{y} - \mathbf{X}\mathbf{b}$ holds for all $\mathbf{y}$.

Summarizing, the *LU* disturbance estimator $\mathbf{B}'\mathbf{y}$ with $\mathbf{B}'\mathbf{X} = \mathbf{0}$ is a regression residual vector if and only if $\mathbf{B}$ is an idempotent $n \times n$ matrix with rank $n$-$k$. In that case $\mathbf{b}$ in $\mathbf{B}'\mathbf{y} = \mathbf{y} - \mathbf{X}\mathbf{b}$ can be computed from $\mathbf{B}$ by $\mathbf{b} = (\mathbf{X}'\mathbf{X})^{-1}\mathbf{X}'(\mathbf{I}-\mathbf{B}')\mathbf{y}$. It is easily verified that $\mathbf{b}$ is a $LU\boldsymbol{\beta}$-estimator. To see that $\mathbf{v} = \mathbf{y} - \mathbf{X}\mathbf{b}$ with $\mathbf{b} = (\mathbf{X}'\mathbf{X})^{-1}\mathbf{X}'(\mathbf{I}-\mathbf{B}')\mathbf{y}$ yields $\mathbf{v} = \mathbf{B}'\mathbf{y}$, consider $\mathbf{v} - \mathbf{B}'\mathbf{y} = \mathbf{y} - \mathbf{B}'\mathbf{y} - \mathbf{X}\mathbf{b} = (\mathbf{I}-\mathbf{B}')\mathbf{y} - \mathbf{X}(\mathbf{X}'\mathbf{X})^{-1}\mathbf{X}'(\mathbf{I}-\mathbf{B}')\mathbf{y} = \mathbf{M}(\mathbf{I}-\mathbf{B}')\mathbf{y} = \mathbf{o}$, where $\mathbf{M}(\mathbf{I}-\mathbf{B}') = \mathbf{0}$ follows from $\mathcal{M}(\mathbf{I}-\mathbf{B}') = \mathcal{M}(\mathbf{X}) = \mathcal{M}(\mathbf{M})^{\perp}$.

Generally, a *BLUF* disturbance estimator is not a regression residual vector, since $\mathbf{B}$ in the *BLUF* estimator $\mathbf{w} = \mathbf{B}'\mathbf{y}$ is generally not idempotent. To find the *BLUF* estimators which are residual vectors, we impose the property $\mathbf{B} = \mathbf{B}^2$ on all $n \times n$ matrices $\mathbf{B}'$ with rank $n$-$k$ which can be written as $\mathbf{K}\mathbf{H}'\overline{\mathbf{N}}'\mathbf{S}^{-1}$ with $\mathbf{H}' = \mathbf{H}^{-1}$.

From $\mathbf{B}' = \mathbf{K}\mathbf{H}'\overline{\mathbf{N}}'\mathbf{S}^{-1}$ with $\mathbf{H}' = \mathbf{H}^{-1}$ and $\mathbf{B} = \mathbf{B}^2$ it follows that:

$$\mathbf{H} = \overline{\mathbf{N}}'\mathbf{S}^{-1}\mathbf{K}$$

Substituting this $\mathbf{H}$ into $\mathbf{B}' = \mathbf{K}\mathbf{H}'\overline{\mathbf{N}}'\mathbf{S}^{-1}$ we get:

$$\mathbf{B}' = \Omega\Gamma^{-1}\mathbf{M}^*$$

Hence, $\mathbf{B}$ is completely determined by $\Omega$, $\Gamma$, and $\mathbf{X}$, without using the best criterion (3.1). Besides, one is not free to choose $\Omega$ for given $\Gamma$ and $\mathbf{X}$, since $\mathbf{K}$ must be chosen in such a way that $\mathbf{H} = \overline{\mathbf{N}}'\mathbf{S}^{-1}\mathbf{K}$ is square and orthogonal. With this restriction on $\mathbf{K}$ it is obvious that one cannot speak of an *a priori* fixed covariance matrix $\sigma^2\Omega$.

Since $\overline{\mathbf{N}}$ and $\mathbf{S}^{-1}\mathbf{X}$ together span the whole $n$-dimensional space, $\mathbf{S}^{-1}\mathbf{K}$ can always be written as $\overline{\mathbf{N}}\mathbf{T}' + \mathbf{S}^{-1}\mathbf{X}\mathbf{G}'$, where $\mathbf{T}$ and $\mathbf{G}$ are unique $(n$-$k) \times (n$-$k)$ and $(n$-$k) \times k$ matrices, respectively, given $\overline{\mathbf{N}}$ and $\mathbf{S}$. It follows that:

$$\mathbf{H} = \overline{\mathbf{N}}'\mathbf{S}^{-1}\mathbf{K} = \overline{\mathbf{N}}'(\overline{\mathbf{N}}\mathbf{T}' + \mathbf{S}^{-1}\mathbf{X}\mathbf{G}') = \mathbf{T}'$$

so that $\mathbf{H}' = \mathbf{H}^{-1}$ if and only if $\mathbf{T}' = \mathbf{T}^{-1}$. In particular, if $\Gamma = \mathbf{I}$ and $\Omega$ is idempotent, which is the case in all our applications, then $\mathbf{K} = \mathbf{N}\mathbf{T}' + \mathbf{X}\mathbf{G}'$ and $\mathbf{K}'\mathbf{K} = \mathbf{I}$. The latter equations imply $\mathbf{G} = \mathbf{0}$, so that $\Omega = \mathbf{M}$ and $\mathbf{B}' = \mathbf{M}$, i.e. the only $\mathbf{w}$, which is a residual vector, is $\hat{\mathbf{u}}$.

To avoid confusion we remark that $\hat{\mathbf{u}}$ is not the only residual vector when $\Gamma = \mathbf{I}$. For instance, if $\mathbf{A}$ is an $n \times n$ matrix such that

$(\mathbf{X}'\mathbf{AX})^{-1}$ exists, then we may take $\mathbf{b} = (\mathbf{X}'\mathbf{AX})^{-1}\mathbf{X}'\mathbf{Ay}$ and $\mathbf{v} = [\mathbf{I} - \mathbf{X}(\mathbf{X}'\mathbf{AX})^{-1}\mathbf{X}'\mathbf{A}]\mathbf{y}$. In this case $\mathbf{v} = \mathbf{y} - \mathbf{Xb}$ is an identity with respect to $\mathbf{y}$, so that $\mathbf{v}$ is a residual vector. The point is that $\hat{\mathbf{u}}$ is the only *LU* disturbance estimator whose covariance matrix (apart from $\sigma^2$) is idempotent.

## 3.5 Durbin's alternative disturbance estimator

For the case $\Gamma = \mathbf{I}_{(n)}$ and $\Omega = \Omega^2_{(n)}$, $rank(\Omega) = n\text{-}k$, Durbin (1970) constructed an alternative *LUF* disturbance estimator $\mathbf{z}$. In his applications Durbin takes $\Omega = \mathbf{KK}' = \mathbf{I} - \mathbf{PP}'$, with $[\mathbf{P} \vdots \mathbf{K}] = \mathbf{H}^*$; see (2.24). Considering only $\mathbf{X}$-matrices including a constant term column, $\mathbf{h}_1^*$, we partition $\mathbf{X}$ and $\mathbf{P}$ as $[\mathbf{h}_1^* \vdots \widetilde{\mathbf{X}}]$ and $[\mathbf{h}_1^* \vdots \mathbf{L}]$, respectively. Durbin proposed the following computing procedure. Let $a$, $\mathbf{b}_1$ and $\mathbf{b}_2$ be the coefficients of $\mathbf{h}_1^*$, $\widetilde{\mathbf{X}}$, and $\mathbf{L}$ in the least-squares fit of the regression of $\mathbf{y}$ on $\mathbf{h}_1^*$, $\widetilde{\mathbf{X}}$ and $\mathbf{L}$. Let $\sigma^2\mathbf{P}_1\mathbf{P}_1'$ and $\sigma^2\mathbf{P}_2\mathbf{P}_2'$ be the covariance matrices of $\mathbf{b}_1$ and $\mathbf{b}_2$, respectively, $\mathbf{P}_1$ and $\mathbf{P}_2$ both being lower triangular matrices. Then $\mathbf{z}$ is computed as:[1]

$$\mathbf{z} = \mathbf{y} - \mathbf{h}_1^* a - \widetilde{\mathbf{X}}\mathbf{b}_1 - \mathbf{L}\mathbf{b}_2 + \Omega\widetilde{\mathbf{X}}\mathbf{P}_1\mathbf{P}_2^{-1}\mathbf{b}_2 \tag{3.3}$$

Given $\Gamma = \mathbf{I}_{(n)}$ and $rank(\Omega) = n\text{-}k$, every *LUF* disturbance estimator can be written as $\mathbf{KH}'\mathbf{N}'\mathbf{y}$ with $\mathbf{H}' = \mathbf{H}^{-1}$. Below we show that:

$$\mathbf{z} = \mathbf{KH}_d'\mathbf{N}'\mathbf{y} \tag{3.4}$$

where:

$$\mathbf{H}_d^{-1} = \mathbf{H}_d' = [\mathbf{I}_{(n\text{-}k)} - \mathbf{K}'\widetilde{\mathbf{X}}(\widetilde{\mathbf{X}}'\Omega\widetilde{\mathbf{X}})^{-1}\widetilde{\mathbf{X}}'\mathbf{K}]\mathbf{K}'\mathbf{N} + \mathbf{K}'\widetilde{\mathbf{X}}\mathbf{P}_1\mathbf{P}_2'\mathbf{L}'\mathbf{N}$$

The lower triangular forms of $\mathbf{P}_1$ and $\mathbf{P}_2$ are not necessary to obtain $\mathcal{V}(\mathbf{z}) = \sigma^2\Omega$; instead, it is sufficient that the product matrix $\mathbf{P}_1\mathbf{P}_2'$ satisfies $\mathbf{P}_1\mathbf{P}_2'\mathbf{L}'\mathbf{MLP}_2\mathbf{P}_1' = (\widetilde{\mathbf{X}}'\Omega\widetilde{\mathbf{X}})^{-1}$. We prove that a specification of $\mathbf{P}_1\mathbf{P}_2'$ exists, such that, when substituted into (3.4), a vector $\mathbf{z}_t$ results, which is equal to the *BLUF* estimator $\mathbf{w} = \mathbf{K}(\mathbf{K}'\mathbf{MK})^{-1/2}\mathbf{K}'\mathbf{My}$, provided $\mathbf{w}$ exists.

1. Recently, Sims (1975) proposed replacing $\mathbf{L}$ by $\mathbf{LL}'\widetilde{\mathbf{X}}$ in the least-squares fit, and reversing the sign before $\Omega$ in (3.3). See also Dubbelman et al. (1976, Appendix B).

For the derivation of (3.4) we need expressions for $a$, $\mathbf{b}_1$, and $\mathbf{b}_2$. Let $\hat{\boldsymbol{\gamma}} = [a \vdots \mathbf{b}_1' \vdots \mathbf{b}_2']' = (\mathbf{Z}'\mathbf{Z})^{-1}\mathbf{Z}'\mathbf{y}$, where:

$$\mathbf{Z} = [\mathbf{h} \vdots \tilde{\mathbf{X}} \vdots \mathbf{L}] = [\mathbf{X} \vdots \mathbf{L}]$$

Here we write $\mathbf{h}$ instead of $\mathbf{h}_1^*$, for notational convenience. It can be verified that :

$$\mathbf{Z}'\mathbf{Z} = \begin{bmatrix} 1 & \mathbf{h}'\tilde{\mathbf{X}} & \mathbf{o}' \\ \tilde{\mathbf{X}}'\mathbf{h} & \tilde{\mathbf{X}}'\tilde{\mathbf{X}} & \tilde{\mathbf{X}}'\mathbf{L} \\ \mathbf{o} & \mathbf{L}'\tilde{\mathbf{X}} & \mathbf{I} \end{bmatrix} = \begin{bmatrix} \mathbf{X}'\mathbf{X} & \mathbf{X}'\mathbf{L} \\ \mathbf{L}'\mathbf{X} & \mathbf{I} \end{bmatrix}$$

and that :

$$(\mathbf{Z}'\mathbf{Z})^{-1} = \begin{bmatrix} 1 + \mathbf{h}'\tilde{\mathbf{X}}\mathbf{A}\tilde{\mathbf{X}}'\mathbf{h} & -\mathbf{h}'\tilde{\mathbf{X}}\mathbf{A} & \mathbf{h}'\tilde{\mathbf{X}}\mathbf{A}\tilde{\mathbf{X}}'\mathbf{L} \\ -\mathbf{A}\tilde{\mathbf{X}}'\mathbf{h} & \mathbf{A} & -\mathbf{A}\tilde{\mathbf{X}}'\mathbf{L} \\ \mathbf{L}'\tilde{\mathbf{X}}\mathbf{A}\tilde{\mathbf{X}}'\mathbf{h} & -\mathbf{L}'\tilde{\mathbf{X}}\mathbf{A} & \mathbf{I} + \mathbf{L}'\tilde{\mathbf{X}}\mathbf{A}\tilde{\mathbf{X}}'\mathbf{L} \end{bmatrix}$$

$$= \begin{bmatrix} (\mathbf{X}'\mathbf{X})^{-1} + \mathbf{G}'(\mathbf{L}'\mathbf{M}\mathbf{L})^{-1}\mathbf{G} & -\mathbf{G}'(\mathbf{L}'\mathbf{M}\mathbf{L})^{-1} \\ -(\mathbf{L}'\mathbf{M}\mathbf{L})^{-1}\mathbf{G} & (\mathbf{L}'\mathbf{M}\mathbf{L})^{-1} \end{bmatrix}$$

with $\mathbf{A} = (\tilde{\mathbf{X}}'\Omega\tilde{\mathbf{X}})^{-1}$ and $\mathbf{G} = \mathbf{L}'\mathbf{X}(\mathbf{X}'\mathbf{X})^{-1}$. Note that $\Omega = \mathbf{K}\mathbf{K}' = \mathbf{I} - \mathbf{P}\mathbf{P}' = \mathbf{I} - \mathbf{h}\mathbf{h}' - \mathbf{L}\mathbf{L}'$. We find :

$$\mathbf{P}_1\mathbf{P}_1' = (\widetilde{\mathbf{X}}'\Omega\widetilde{\mathbf{X}})^{-1}$$

$$\mathbf{P}_2\mathbf{P}_2' = \mathbf{I} + \mathbf{L}'\widetilde{\mathbf{X}}\mathbf{A}\widetilde{\mathbf{X}}'\mathbf{L} = (\mathbf{L}'\mathbf{M}\mathbf{L})^{-1}$$

and, from the first partitioning of **Z**:

$$\hat{\boldsymbol{\gamma}} = \begin{bmatrix} a \\ \cdots \\ \mathbf{b}_1 \\ \cdots \\ \mathbf{b}_2 \end{bmatrix} = \begin{bmatrix} \mathbf{h}'(\mathbf{I} + \widetilde{\mathbf{X}}\mathbf{A}\widetilde{\mathbf{X}}'\mathbf{hh}' - \widetilde{\mathbf{X}}\mathbf{A}\widetilde{\mathbf{X}}' + \widetilde{\mathbf{X}}\mathbf{A}\widetilde{\mathbf{X}}'\mathbf{LL}') \\ \cdots \\ -\mathbf{A}\widetilde{\mathbf{X}}'\mathbf{hh}' + \mathbf{A}\widetilde{\mathbf{X}}' - \mathbf{A}\widetilde{\mathbf{X}}'\mathbf{LL}' \\ \cdots \\ \mathbf{L}'(\widetilde{\mathbf{X}}\mathbf{A}\widetilde{\mathbf{X}}'\mathbf{hh}' - \widetilde{\mathbf{X}}\mathbf{A}\widetilde{\mathbf{X}}' + \mathbf{I} + \widetilde{\mathbf{X}}\mathbf{A}\widetilde{\mathbf{X}}'\mathbf{LL}') \end{bmatrix} \mathbf{y}$$

$$= \begin{bmatrix} \mathbf{h}'(\mathbf{I} - \widetilde{\mathbf{X}}\mathbf{A}\widetilde{\mathbf{X}}'\Omega)\mathbf{y} \\ \cdots \\ \mathbf{A}\widetilde{\mathbf{X}}'\Omega\mathbf{y} \\ \cdots \\ \mathbf{L}'(\mathbf{I} - \widetilde{\mathbf{X}}\mathbf{A}\widetilde{\mathbf{X}}'\Omega)\mathbf{y} \end{bmatrix}$$

From the second partitioning of **Z** it follows that $\mathbf{b}_2 = (\mathbf{L}'\mathbf{M}\mathbf{L})^{-1}\mathbf{L}'\mathbf{M}\mathbf{y}$, so that $\mathbf{P}_2^{-1}\mathbf{b}_2 = \mathbf{P}_2^{-1}\mathbf{P}_2\mathbf{P}_2'\mathbf{L}'\mathbf{M}\mathbf{y} = \mathbf{P}_2'\mathbf{L}'\mathbf{M}\mathbf{y}$. Substition of the expressions for $a$, $\mathbf{b}_1$, $\mathbf{b}_2$, and $\mathbf{P}_2^{-1}\mathbf{b}_2$ into (3.3) yields :

$$\begin{aligned} \mathbf{z} &= [\mathbf{I} - \mathbf{hh}'(\mathbf{I} - \widetilde{\mathbf{X}}\mathbf{A}\widetilde{\mathbf{X}}'\Omega) - \widetilde{\mathbf{X}}\mathbf{A}\widetilde{\mathbf{X}}'\Omega - \mathbf{LL}'(\mathbf{I} - \widetilde{\mathbf{X}}\mathbf{A}\widetilde{\mathbf{X}}'\Omega) \\ &\quad + \Omega\widetilde{\mathbf{X}}\mathbf{P}_1\mathbf{P}_2'\mathbf{L}'\mathbf{M}]\mathbf{y} \\ &= [\mathbf{I} - \mathbf{hh}' - \mathbf{LL}' - (\mathbf{I} - \mathbf{hh}' - \mathbf{LL}')\widetilde{\mathbf{X}}\mathbf{A}\widetilde{\mathbf{X}}'\Omega + \Omega\widetilde{\mathbf{X}}\mathbf{P}_1\mathbf{P}_2'\mathbf{L}'\mathbf{M}]\mathbf{y} \\ &= \Omega\,[\mathbf{I} - \widetilde{\mathbf{X}}\mathbf{A}\widetilde{\mathbf{X}}'\Omega + \widetilde{\mathbf{X}}\mathbf{P}_1\mathbf{P}_2'\mathbf{L}'\mathbf{M}]\mathbf{y} \\ &= \mathbf{K}(\mathbf{V}\mathbf{K}' + \mathbf{U}\mathbf{M})\mathbf{y} \end{aligned}$$

where :

$$\mathbf{V} = \mathbf{I}_{(n-k)} - \mathbf{K}'\widetilde{\mathbf{X}}(\widetilde{\mathbf{X}}'\Omega\widetilde{\mathbf{X}})^{-1}\widetilde{\mathbf{X}}'\mathbf{K}$$

$$\mathbf{U} = \mathbf{K}'\widetilde{\mathbf{X}}\mathbf{P}_1\mathbf{P}_2'\mathbf{L}'$$

Note that $\mathbf{V} = \mathbf{V}' = \mathbf{V}^2$, $\mathbf{VK}'\mathbf{X} = \mathbf{V}[\mathbf{K}'\mathbf{h} \vdots \mathbf{K}'\widetilde{\mathbf{X}}] = \mathbf{V}[\mathbf{o} \vdots \mathbf{K}'\widetilde{\mathbf{X}}] = \mathbf{0}$ and hence $\mathbf{VK}'\mathbf{M} = \mathbf{VK}'$. Further, $\mathbf{UK} = \mathbf{0}$ since $\mathbf{L}'\mathbf{K} = \mathbf{0}$, and $\mathbf{UMU}' = \mathbf{K}'\widetilde{\mathbf{X}}\mathbf{P}_1\mathbf{P}_2'\mathbf{L}'\mathbf{MLP}_2\mathbf{P}_1'\widetilde{\mathbf{X}}'\mathbf{K} = \mathbf{K}'\widetilde{\mathbf{X}}\mathbf{P}_1\mathbf{P}_1'\widetilde{\mathbf{X}}'\mathbf{K} = \mathbf{I} - \mathbf{V}$. Now consider $\mathbf{H}_d' = \mathbf{VK}'\mathbf{N} + \mathbf{UN}$. Then :

$$\mathbf{KH}_d'\mathbf{N}'\mathbf{y} = \mathbf{K}(\mathbf{VK}'\mathbf{NN}' + \mathbf{UNN}')\mathbf{y} = \mathbf{K}(\mathbf{VK}'\mathbf{M} + \mathbf{UM})\mathbf{y} = \mathbf{z}$$

and :

$$\begin{aligned}\mathbf{H}_d'\mathbf{H}_d &= \mathbf{VK}'\mathbf{MKV} + \mathbf{VK}'\mathbf{MU}' + \mathbf{UMKV} + \mathbf{UMU}' \\ &= \mathbf{V} + \mathbf{0} + \mathbf{0} + \mathbf{I} - \mathbf{V} = \mathbf{I}\end{aligned}$$

which proves (3.4).

From the above derivation it is clear that it is not at all necessary that $\mathbf{P}_1$ and $\mathbf{P}_2$ are lower triangular matrices in order that $\mathbf{z}$ be a *LUF* disturbance estimator. Below we consider the vector $\mathbf{z}_t$ :

$$\mathbf{z}_t = \mathbf{KH}_t'\mathbf{N}'\mathbf{y}$$

where :

$$\mathbf{H}_t' = \mathbf{VK}'\mathbf{N} + \mathbf{U}_t\mathbf{N}$$

and :

$$\mathbf{U}_t = \mathbf{K}'\widetilde{\mathbf{X}}(\widetilde{\mathbf{X}}'\Omega\widetilde{\mathbf{X}})^{-1/2}\mathbf{T}'(\mathbf{L}'\mathbf{ML})^{-1/2}\mathbf{L}'$$

Note that $\mathbf{z}_t = \mathbf{z}$ for $\mathbf{T}' = (\widetilde{\mathbf{X}}'\Omega\widetilde{\mathbf{X}})^{1/2}\mathbf{P}_1\mathbf{P}_2'(\mathbf{L}'\mathbf{ML})^{1/2}$.

We wish to answer the following question: is $\mathbf{z}$ equal to $\mathbf{w}$, and if not, does a matrix $\mathbf{T}$ exist such that $\mathbf{z}_t = \mathbf{w} = \mathbf{K}(\mathbf{K}'\mathbf{MK})^{-1/2}\mathbf{K}'\mathbf{My}$? The vector $\mathbf{w}$ is that vector $\mathbf{v}$ which uniquely minimizes the objective function :

$$\epsilon(\mathbf{v}) = \frac{1}{\sigma^2}\,\mathcal{E}[(\mathbf{v} - \mathbf{u})'(\mathbf{v} - \mathbf{u})] = 2n - k - 2tr(\mathbf{B})$$

where $\mathbf{v} = \mathbf{B}'\mathbf{y}$ with $\mathbf{B}'\mathbf{X} = \mathbf{0}$ and $\mathbf{B}'\mathbf{B} = \Omega$. Here $\Omega = \Omega' = \Omega^2$ and $rank(\Omega) = n\text{-}k$. Hence :

$$\epsilon(\mathbf{w}) = 2n - k - 2tr(\mathbf{K}'\mathbf{MK})^{1/2}$$

It is proved in the appendix to Chapter 4, that:

$$tr(\mathbf{K}'\mathbf{M}\mathbf{K})^{1/2} = n - 2k + \sum_{i=1}^{k} d_i^{1/2}$$

where the $d_i$'s are the eigenvalues of $\mathbf{P}'(\mathbf{I}-\mathbf{M})\mathbf{P}$:

$$\mathbf{P}'(\mathbf{I}-\mathbf{M})\mathbf{P} = \mathbf{I} - [\mathbf{h} \vdots \mathbf{L}]'\mathbf{M}[\mathbf{h} \vdots \mathbf{L}] = \begin{bmatrix} 1 & \vdots & \mathbf{o}' \\ \cdots & \vdots & \cdots \\ \mathbf{o} & \vdots & \mathbf{L}'(\mathbf{I}-\mathbf{M})\mathbf{L} \end{bmatrix}$$

Writing $\mathbf{L}'(\mathbf{I}-\mathbf{M})\mathbf{L} = \mathbf{F}^*\mathbf{D}^*\mathbf{F}^{*\prime}$ with $\mathbf{F}^{*\prime} = \mathbf{F}^{*-1}$, we have $\mathbf{P}'(\mathbf{I}-\mathbf{M})\mathbf{P} = \mathbf{F}\mathbf{D}\mathbf{F}'$ with $\mathbf{F}' = \mathbf{F}^{-1}$ and:

$$\mathbf{F} = \begin{bmatrix} 1 & \vdots & \mathbf{o}' \\ \cdots & \vdots & \cdots \\ \mathbf{o} & \vdots & \mathbf{F}^* \end{bmatrix} \text{ and } \mathbf{D} = \begin{bmatrix} 1 & \vdots & \mathbf{o}' \\ \cdots & \vdots & \cdots \\ \mathbf{o} & \vdots & \mathbf{D}^* \end{bmatrix}$$

Hence, $\sum_{i=1}^{k} d_i^{1/2} = tr(\mathbf{D}^{1/2}) = 1 + tr(\mathbf{D}^{*1/2}) = 1 + tr(\mathbf{F}^*\mathbf{D}^{*1/2}\mathbf{F}^{*\prime}) = 1 + tr[\mathbf{L}'(\mathbf{I}-\mathbf{M})\mathbf{L}]^{1/2}$, so that we find:

$$\begin{aligned} \epsilon(\mathbf{w}) &= 2n - k - 2\,(n - 2k + 1 + tr[\mathbf{L}'(\mathbf{I}-\mathbf{M})\mathbf{L}]^{1/2}) \\ &= 3k - 2 - 2tr[\mathbf{L}'(\mathbf{I}-\mathbf{M})\mathbf{L}]^{1/2} \end{aligned}$$

For several **X**-matrices, described in Section 5.1, we calculated $\epsilon(\mathbf{w})$ and $\epsilon(\mathbf{z}) = 2n - k - 2tr(\mathbf{K}\mathbf{H}_d'\mathbf{N}')$ – see Table 3.1. In all four cases we find $\mathbf{z} \neq \mathbf{w}$ since $\epsilon(\mathbf{z}) \neq \epsilon(\mathbf{w})$.

*Table 3.1.* Objective function values of w and z

| | $X_C$ | $X_K$ | $X_S$ | $X_T$ |
|---|---|---|---|---|
| $\epsilon$(w) | 3.48 | 5.05 | 4.59 | 3.36 |
| $\epsilon$(z) | 6.97 | 5.60 | 4.61 | 6.47 |

The vector $\mathbf{z}_t$ is a *LUF* disturbance estimator if and only if $\mathbf{H}_t$ is orthogonal. Using $\mathbf{VK}'\mathbf{MU}_t' = \mathbf{VK}'\mathbf{U}_t' = \mathbf{0}$ and:

$$\begin{aligned}\mathbf{U}_t\mathbf{MU}_t' = & \ \mathbf{K}'\tilde{\mathbf{X}}(\tilde{\mathbf{X}}'\Omega\tilde{\mathbf{X}})^{-\frac{1}{2}}\mathbf{T}'(\mathbf{L}'\mathbf{ML})^{-\frac{1}{2}}\mathbf{L}'\mathbf{ML}(\mathbf{L}'\mathbf{ML})^{-\frac{1}{2}} \\ & \ \mathbf{T}(\tilde{\mathbf{X}}'\Omega\mathbf{X})^{-\frac{1}{2}}\tilde{\mathbf{X}}'\mathbf{K} \\ = & \ \mathbf{K}'\tilde{\mathbf{X}}(\tilde{\mathbf{X}}'\Omega\tilde{\mathbf{X}})^{-\frac{1}{2}}\mathbf{T}'\mathbf{T}(\tilde{\mathbf{X}}'\Omega\tilde{\mathbf{X}})^{-\frac{1}{2}}\tilde{\mathbf{X}}'\mathbf{K}\end{aligned}$$

it follows from:

$$\mathbf{I} = \mathbf{H}_t'\mathbf{H}_t = \mathbf{V} + \mathbf{0} + \mathbf{0} + \mathbf{K}'\tilde{\mathbf{X}}(\tilde{\mathbf{X}}'\Omega\tilde{\mathbf{X}})^{-\frac{1}{2}}\mathbf{T}'\mathbf{T}(\tilde{\mathbf{X}}'\Omega\tilde{\mathbf{X}})^{-\frac{1}{2}}\tilde{\mathbf{X}}'\mathbf{K}$$

that $\mathbf{T}$ is orthogonal. Actually, $\mathbf{z}_t$ defines a class of vectors, one vector for each orthogonal $\mathbf{T}$. Because $\epsilon(\mathbf{v})$ is *uniquely* minimized for $\mathbf{v} = \mathbf{w}$, we have $\mathbf{z}_t = \mathbf{w}$ if and only if $\epsilon(\mathbf{z}_t) = \epsilon(\mathbf{w}) = 3k - 2 - 2tr[\mathbf{L}'(\mathbf{I} - \mathbf{M})\mathbf{L}]$. We now determine the minimum value of $\epsilon(\mathbf{z}_t)$ with respect to $\mathbf{T}$ subject to $\mathbf{T}'\mathbf{T} = \mathbf{I}$.

$$\begin{aligned}\epsilon(\mathbf{z}_t) & = 2n - k - 2tr(\mathbf{KH}_t'\mathbf{N}') \\ & = 2n - k - 2tr(\mathbf{KVK}' + \mathbf{KU}_t\mathbf{M}) \\ & = 2n - k - 2tr(\mathbf{V}) - 2tr[\Omega\tilde{\mathbf{X}}(\tilde{\mathbf{X}}'\Omega\tilde{\mathbf{X}})^{-\frac{1}{2}}\mathbf{T}'(\mathbf{L}'\mathbf{ML})^{-\frac{1}{2}}\mathbf{L} \\ & = 2n - k - 2[(n-k) - (k-1)] - 2tr(\mathbf{T}'\mathbf{W}) \\ & = 3k - 2 - 2tr(\mathbf{T}'\mathbf{W})\end{aligned}$$

where $\mathbf{W} = (\mathbf{L}'\mathbf{ML})^{-\frac{1}{2}}\mathbf{L}'\mathbf{M}\Omega\tilde{\mathbf{X}}(\mathbf{X}'\Omega\tilde{\mathbf{X}})^{-\frac{1}{2}}$. In view of $\mathbf{M}\Omega\tilde{\mathbf{X}} = \mathbf{M}(\mathbf{I} - \mathbf{hh}$ $- \mathbf{LL}')\tilde{\mathbf{X}} = (-\mathbf{MLL}'\tilde{\mathbf{X}})$, we have:

$$\mathbf{W} = -(\mathbf{L}'\mathbf{ML})^{\frac{1}{2}}\mathbf{L}'\tilde{\mathbf{X}}(\tilde{\mathbf{X}}'\Omega\tilde{\mathbf{X}})^{-\frac{1}{2}}$$

In accordance with Result 1.A.6.8, the maximum of $tr(\mathbf{T}'\mathbf{W})$ subject to $\mathbf{T}'\mathbf{T} = \mathbf{I}$ is $tr(\mathbf{W}'\mathbf{W})^{\frac{1}{2}}$:

$$\begin{aligned}tr(\mathbf{W}'\mathbf{W})^{\frac{1}{2}} & = tr(\mathbf{WW}') = tr[(\mathbf{L}'\mathbf{ML})^{\frac{1}{2}}\mathbf{L}'\tilde{\mathbf{X}}(\tilde{\mathbf{X}}'\Omega\tilde{\mathbf{X}})^{-1}\tilde{\mathbf{X}}'\mathbf{L}(\mathbf{L}'\mathbf{ML} \\ & = tr[(\mathbf{L}'\mathbf{ML})^{\frac{1}{2}}\left\{(\mathbf{L}'\mathbf{ML})^{-1} - \mathbf{I}\right\}(\mathbf{L}'\mathbf{ML})^{\frac{1}{2}}]^{\frac{1}{2}} \\ & = tr(\mathbf{I} - \mathbf{L}'\mathbf{ML})^{\frac{1}{2}} = tr[\mathbf{L}'(\mathbf{I} - \mathbf{M})\mathbf{L}]^{\frac{1}{2}}\end{aligned}$$

so that the minimum value of $\epsilon(\mathbf{z}_t)$ is:

$$\epsilon(\mathbf{z}_t) = 3k - 2 - 2tr[\mathbf{L}'(\mathbf{I} - \mathbf{M})\mathbf{L}]^{1/2} = \epsilon(\mathbf{w})$$

When $k = 2$, the matrices $\mathbf{T}$ and $\mathbf{W}$ are scalars. Hence, $\mathbf{T} = 1$ or $\mathbf{T} = -1$. Remember that $\mathbf{z}_t = \mathbf{z}$ for $\mathbf{T} = (\tilde{\mathbf{X}}'\Omega\tilde{\mathbf{X}})^{1/2}\mathbf{P}_1\mathbf{P}_2'(\mathbf{L}'\mathbf{M}\mathbf{L})^{1/2}$, so that $\mathbf{T} = 1$. But minimization of $\epsilon(\mathbf{z}_t) = 3k - 2 - 2tr(\mathbf{T}'\mathbf{W})$ subject to $\mathbf{T}'\mathbf{T} = 1$ requires that $\mathbf{T}$ and $\mathbf{W}$ have the same sign. From $\mathbf{W} = -(\mathbf{L}'\mathbf{M}\mathbf{L})^{1/2}\mathbf{L}'\tilde{\mathbf{X}}(\tilde{\mathbf{X}}'\Omega\tilde{\mathbf{X}})^{-1/2}$ it follows that $\mathbf{W}$ and $\mathbf{L}'\tilde{\mathbf{X}}$ have opposite sign. Hence, when the scalar $\mathbf{L}'\tilde{\mathbf{X}}$ is negative, then $\mathbf{z}_t = \mathbf{w}$ for $\mathbf{T} = 1$, and when $\mathbf{L}'\tilde{\mathbf{X}}$ is positive, then $\mathbf{z}_t = \mathbf{w}$ for $\mathbf{T} = -1$.

# 4. An empirical Ω

## 4.1 From the general to a specific w

We hope to arrive at powerful tests using test statistics of the quadratic ratio type (2.1) with significance points that can be tabulated. For this purpose, we developed the *BLUF* estimator **w** of $\mathbf{J}'\mathbf{u}$; see (3.2). The estimator depends on the following matrices: the $n \times p$ matrix **J**, the $p \times p$ matrix **Q**, the $p \times p$ matrix $\Omega = \mathbf{KK}'$, the $n \times n$ matrix Γ, the $n \times k$ matrix **X**, and the $n$-element vector **y**. Both **X** and **y** are specified by observation and it is assumed that $\mathbf{y} \sim \mathcal{N}(\mathbf{X}\boldsymbol{\beta}, \sigma^2\Gamma)$, where $\Gamma = \Gamma_0$ under the null hypothesis $\mathcal{H}_0$. Hence, Γ in (3.2) is specified by $\mathcal{H}_0$. One is free to choose **J** and **Q**. The only practical specifications we know are $\mathbf{J} = \mathbf{I}_{(n)}$ or **J** is an $n \times (n\text{-}k)$ submatrix of $\mathbf{I}_{(n)}$, like **J** in the *BLUS* vectors, and $\mathbf{Q} = \mathbf{I}$ or $\mathbf{Q} = \Gamma^{-1}$. In this chapter we take $\mathbf{J} = \mathbf{I}$, so that $p = n$. Because most current tests have $\Gamma = \mathbf{I}$ as the null hypothesis, the most important specifications for practical application would seem to be $\Gamma = \mathbf{Q} = \mathbf{J} = \mathbf{I}_{(n)}$. In this chapter we therefore restrict ourselves to this specific **w**:

$$\mathbf{w} = \mathbf{B}'\mathbf{y} = \mathbf{K}(\mathbf{K}'\mathbf{M}\mathbf{K})^{-1/2}\mathbf{K}'\mathbf{M}\mathbf{y} \tag{4.1}$$

where the $n \times r$ matrix **K** has rank $r$, $r \leqslant n\text{-}k$. The crucial point is the choice of Ω, which is the subject of this chapter. Of course, we wish to choose Ω such that the test is as powerful as possible. In Chapter 2 we have seen that the power of a test is sometimes maximized for $\mathbf{w} = \hat{\mathbf{u}}$, so that $\Omega = \mathbf{M}$. However, in such cases Ω would depend on **X**, and we wish to choose Ω prior to the specification of **X**. We are not able to solve the problem of power maximization under the restriction of an *a priori* fixed covariance matrix Ω. Instead, we develop a criterion for Ω in terms of estimation theory, which is done in Section

4.2. According to this criterion $\Omega$ must be a mean **M**-matrix. The method of obtaining this $\Omega$ from a set of empirical **X**-matrices is discussed in Section 4.3. In Section 4.4 the method is applied to economic time series **X**-matrices, and the resulting $\Omega$ is streamlined and generalized in Sections 4.5 and 4.6. Of course, alternative choices of $\Omega$ remain possible. One alternative is proposed in Section 4.7. Another alternative is to choose $\Omega$ such that the elements of **w** are mutually stochastically independent, like the elements of a *BLUS* vector. In Chapter 5 the various $\Omega$'s are evaluated in terms of powers.

## 4.2 Measures for $\Omega$

In Chapter 3 we assumed that $\Omega$ is a given matrix. Then the vector **w** in (4.1) is the best of all *LUF* estimators of **u** (and of $\hat{\mathbf{u}}$), best in the sense that $\epsilon(\mathbf{v})$ is minimized for $\mathbf{v} = \mathbf{w}$:

$$\epsilon(\mathbf{v}) = \frac{1}{\sigma^2}\, \mathcal{E}[(\mathbf{v} - \mathbf{u})'(\mathbf{v} - \mathbf{u})]$$

Suppose that we have two $\Omega$-matrices, say $\Omega_1 = \mathbf{K}_1\mathbf{K}_1'$ and $\Omega_2 = \mathbf{K}_2\mathbf{K}_2'$. Then two **w**'s can be computed, say $\mathbf{w}_1 = \mathbf{w}(\Omega_1)$ and $\mathbf{w}_2 = \mathbf{w}(\Omega_2)$. According to the estimation criterion, $\mathbf{w}_1$ is better than $\mathbf{w}_2$ if $\epsilon(\mathbf{w}_1) < \epsilon(\mathbf{w}_2)$. In that case we may also say that $\Omega_1$ is better than $\Omega_2$, since the difference between $\mathbf{w}_1$ and $\mathbf{w}_2$ is caused by the difference between $\Omega_1$ and $\Omega_2$. Hence, $\epsilon(\mathbf{w})$ provides us with a measure for $\Omega$:

$$\begin{aligned}\epsilon(\mathbf{w}) &= \frac{1}{\sigma^2}\, \mathcal{E}[\mathbf{u}'(\mathbf{B} - \mathbf{I})(\mathbf{B}' - \mathbf{I})\mathbf{u}] \\ &= tr(\mathbf{BB}' - \mathbf{B} - \mathbf{B}' + \mathbf{I}) \\ &= n + tr(\Omega) - 2tr(\mathbf{B}) \\ &= n + tr(\mathbf{K}'\mathbf{K}) - 2tr(\mathbf{K}'\mathbf{MK})^{1/2}\end{aligned}$$

Minimization of $\epsilon(\mathbf{w})$ with respect to $\Omega$ would yield $\Omega = \mathbf{M}$, since the *BLUF* estimation problem reduces to the *BLU* estimation problem if $\Omega$ is not fixed. Therefore, $\epsilon(\mathbf{w})$ is to be used as an appreciation measure rather than as an optimization criterion. In the next paragraph we show that $\epsilon(\mathbf{w})$ is not yet a good measure for $\Omega$.

Suppose $\mathbf{K}_2 = c\mathbf{K}_1$, where $c$ is an arbitrary positive scalar. Then $\mathbf{w}_2 = \mathbf{K}_2(\mathbf{K}_2'\mathbf{M}\mathbf{K}_2)^{-1/2}\mathbf{K}_2'\mathbf{M}\mathbf{y} = c\mathbf{K}_1(c\mathbf{K}_1'\mathbf{M}c\mathbf{K}_1)^{-1/2}\, c\mathbf{K}_1'\mathbf{M}\mathbf{y} = c\mathbf{K}_1(\mathbf{K}_1'\mathbf{M}\mathbf{K}_1$ $\mathbf{K}_1'\mathbf{M}\mathbf{y} = c\mathbf{w}_1$. The test statistic $T(\mathbf{w}) = \mathbf{w}'\mathbf{A}\mathbf{w}/\mathbf{w}'\mathbf{C}\mathbf{w}$ is invariant under scalar multiplication of $\mathbf{w}$, i.e. $T(\mathbf{w}_1) = T(\mathbf{w}_2)$. Hence, for testing purposes $\Omega_1$ and $\Omega_2 = c^2\Omega_1$ are equally good for all $c$, and a good measure for $\Omega$ should be invariant under scalar multiplication of $\Omega$. However, $\epsilon[\mathbf{w}(c^2\Omega)] = n + c^2 tr(\mathbf{K}'\mathbf{K}) - 2c\, tr(\mathbf{K}'\mathbf{M}\mathbf{K})^{1/2}$, which shows that $\epsilon(\mathbf{w})$ is not invariant under scalar multiplication of $\Omega$. Clearly, some restriction must be imposed. We restrict ourselves to idempotent $\Omega$-matrices, which we justify in the next paragraph. This restriction excludes scalar multiplication of $\Omega$, so that invariance of $\epsilon(\mathbf{w})$ is irrelevant.

Every $\Omega$ can be written as $\Omega = \mathbf{F}\mathbf{D}\mathbf{F}'$ with $\mathbf{F}$ satisfying $\mathbf{F}'\mathbf{F} = \mathbf{I}_{(r)}$ (see also Section 2.2). Then $\Omega$ can be chosen by means of $\mathbf{F}$ and $\mathbf{D}$. The vector $\mathbf{w}$ lies in $\mathcal{M}(\mathbf{K}) = \mathcal{M}(\mathbf{F})$, while $\hat{\mathbf{u}}$ lies in $\mathcal{M}(\mathbf{M})$. Since $\mathbf{w}$ is the *BLUF* estimator of $\hat{\mathbf{u}}$, it is reasonable to choose $\mathbf{F}$ such that $\mathcal{M}(\mathbf{F})$ is close to $\mathcal{M}(\mathbf{M})$. Moreover, one should choose $r = rank(\mathbf{F})$ as great as possible, because this offers the best opportunities that $\mathbf{w}$ is close to $\hat{\mathbf{u}}$. Since $r \leqslant n\text{-}k$, we take $r = n\text{-}k$ (in Chapter 5 we make some calculations with $r < n\text{-}k$, invariably leading to powers inferior to the case $r = n\text{-}k$). It is to be realized that $\Omega$ is to be applied in connection with several $\mathbf{X}$-matrices, say $\mathbf{X}_1, \mathbf{X}_2, \ldots, \mathbf{X}_m$. In other words, $\mathbf{F}$ is to be chosen such that $\mathcal{M}(\mathbf{F})$ is close to both $\mathcal{M}(\mathbf{M}_1)$ and $\mathcal{M}(\mathbf{M}_2)$ and ... and $\mathcal{M}(\mathbf{M}_m)$. Then it is desirable that $\mathcal{M}(\mathbf{M}_1)$ and $\mathcal{M}(\mathbf{M}_2)$ and ... and $\mathcal{M}(\mathbf{M}_m)$ are close to each other. In that case $\mathbf{F}'\mathbf{X}_1$ and $\mathbf{F}'\mathbf{X}_2$ and ... and $\mathbf{F}'\mathbf{X}_m$ are close to $\mathbf{0}$, so that $\mathbf{F}'\mathbf{M}_1\mathbf{F}$ and $\mathbf{F}'\mathbf{M}_2\mathbf{F}$ and ... and $\mathbf{F}'\mathbf{M}_m\mathbf{F}$ are close to $\mathbf{I}_{(n\text{-}k)}$. Now taking $\mathbf{K} = \mathbf{F}\mathbf{D}^{1/2}$, we have:

$$\begin{aligned}\epsilon(\mathbf{w}) &= n + tr(\mathbf{D}^{1/2}\mathbf{F}'\mathbf{F}\mathbf{D}^{1/2}) - 2tr(\mathbf{D}^{1/2}\mathbf{F}'\mathbf{M}\mathbf{F}\mathbf{D}^{1/2})^{1/2} \\ &= n + tr(\mathbf{D}) - 2tr(\mathbf{D}^{1/2}\mathbf{F}'\mathbf{M}\mathbf{F}\mathbf{D}^{1/2})^{1/2}\end{aligned}$$

which expression is close to:

$$\begin{aligned}n + tr(\mathbf{D}) - 2tr(\mathbf{D}^{1/2}) &= n + \sum_{i=1}^{n\text{-}k} d_i - 2\sum_{i=1}^{n\text{-}k}\sqrt{d_i} \\ &= k + \sum_{i=1}^{n\text{-}k}(\sqrt{d_i} - 1)^2\end{aligned}$$

The latter expression is minimized for $d_1 = d_2 = \ldots = d_{n\text{-}k} = 1$, i.e. $\mathbf{D} = \mathbf{I}_{(n\text{-}k)}$. In that case $\Omega = \mathbf{FF}'$, so that $\Omega = \Omega^2$.

Given $\Omega$ is idempotent, i.e. $\Omega = \mathbf{KK}'$ with $\mathbf{K}'\mathbf{K} = \mathbf{I}_{(n\text{-}k)}$, our appre-cation measure becomes $\epsilon(\mathbf{w}) = n + tr(\mathbf{I}_{(n\text{-}k)}) - 2tr(\mathbf{K}'\mathbf{MK})^{1/2} = 2n - k - 2tr(\mathbf{K}'\mathbf{MK})^{1/2}$. Therefore, for given $\mathbf{M}$, the smaller $tr(\mathbf{K}'\mathbf{MK})^{1/2}$ the better. We recall that $\mathbf{M} = \mathbf{NN}' = \mathbf{I} - \mathbf{RR}'$ with $[\mathbf{R} \vdots \mathbf{N}]' = [\mathbf{R} \vdots \mathbf{N}]^{-1}$ (see Section 1.3). Analogously, when $\Omega$ is an idempotent $n \times n$ matrix with rank $n$-$k$, then $\Omega = \mathbf{KK}' = \mathbf{I} - \mathbf{PP}'$ with $[\mathbf{P} \vdots \mathbf{K}]' = [\mathbf{P} \vdots \mathbf{K}]^{-1}$, $\mathbf{P}$ and $\mathbf{K}$ having the same order as $\mathbf{R}$ and $\mathbf{N}$, respectively. In the appendix to this chapter it is proved that:

$$tr(\mathbf{K}'\mathbf{MK})^{1/2} = n - 2k + tr(\mathbf{P}'\mathbf{RR}'\mathbf{P})^{1/2}$$

Using this result, we define the measure $\psi$:

$$\psi = \frac{1}{k} tr(\mathbf{P}'\mathbf{RR}'\mathbf{P})^{1/2} \qquad (4.2)$$

so that we have, using $\epsilon(\hat{\mathbf{u}}) = k$:

$$\epsilon(\mathbf{w}) - \epsilon(\hat{\mathbf{u}}) = 2k(1 - \psi)$$

As an alternative measure we consider $\phi$:

$$\phi = \frac{1}{k} tr(\mathbf{P}'\mathbf{RR}'\mathbf{P}) \qquad (4.3)$$

which is much easier to calculate than $\psi$. Notice that the sum of squared differences between the elements $m_{ij}$ of $\mathbf{M}$ and the corres-ponding elements $\omega_{ij}$ of $\Omega$ is equal to:

$$\begin{aligned} \sum_{i=1}^{n} \sum_{j=1}^{n} (m_{ij} - \omega_{ij})^2 &= tr[(\mathbf{M} - \Omega)(\mathbf{M} - \Omega)] \\ &= tr[(\mathbf{PP}' - \mathbf{RR}')(\mathbf{PP}' - \mathbf{RR}')] \\ &= 2k(1 - \phi) \end{aligned}$$

Because all eigenvalues of $\mathbf{P}'\mathbf{RR}'\mathbf{P}$ are nonnegative and do not exeed 1, see (4.A.1), we have $0 \leqslant \psi \leqslant 1$ and $0 \leqslant \phi \leqslant 1$. Both $\psi$ and $\phi$ measure the difference between $\Omega$ and $\mathbf{M}$, which matrices are typical of $\mathbf{w}$ and $\hat{\mathbf{u}}$ in the present context; $\psi$ and $\phi$ may be called least-squares approximation measures – the greater the measures, the better the approximation. The measures look very similar, and in fact behave very similarly, as we shall see in Chapter 5. For mathematical convenience we prefer the measure $\phi$. To assist intuition with respect to the measure $\phi$, we make the following remarks. The columns of **R** form an orthonormal basis of the $k$-dimensional regression space $\mathcal{M}(\mathbf{X})$, and the columns of **P** form an orthonormal basis of another $k$-dimensional space. The elements of $\mathbf{P}'\mathbf{R}$ are the cosines of the angles between the column vectors of **P** and the column vectors of **R**, the sum of squared cosines being equal to $tr(\mathbf{P}'\mathbf{R}\mathbf{R}'\mathbf{P}) = k\,\phi$. The measure $\phi$ can also be regarded as the mean of $k$ squared correlation coefficients if one of the columns of **P** is a constant term column, say $\mathbf{p}_1 = (\boldsymbol{\iota}'\boldsymbol{\iota})^{-½}\boldsymbol{\iota}$. In this case we have $\mathbf{p}_i'\mathbf{E}\mathbf{p}_i = \mathbf{p}_i'(\mathbf{I} - \mathbf{p}_1\mathbf{p}_1')\mathbf{p}_i = 1$ for $i = 2, 3, \ldots, k$. Consider the $k$ linear models:

$$\mathbf{p}_i = \mathbf{X}\boldsymbol{\beta}_i + \mathbf{v}_i \qquad i = 1, 2, \ldots, k$$

where **X** contains $\boldsymbol{\iota}$. The squared multiple correlation coefficient in the $i$th model is:

$$R_i^2 = 1 - \frac{\mathbf{p}_i'\mathbf{M}\mathbf{p}_i}{\mathbf{p}_i'\mathbf{E}\mathbf{p}_i} = 1 - \mathbf{p}_i'\mathbf{M}\mathbf{p}_i \qquad i = 2, 3, \ldots, k$$

(see Koerts and Abrahamse 1969, p. 134), and we define $R_1^2 = 1$. Then, using $\mathbf{M}\boldsymbol{\iota} = \mathbf{o}$, we obtain:

$$\frac{1}{k}\sum_{i=1}^{k} R_i^2 = \frac{1}{k}\sum_{i=1}^{k}(1 - \mathbf{p}_i'\mathbf{M}\mathbf{p}_i) = 1 - \frac{1}{k}tr(\mathbf{P}'\mathbf{M}\mathbf{P}) = \phi$$

Now we are able to formulate a criterion to choose Ω on the basis of a set of empirical **X**-matrices. Let $\mathbf{X}_1, \mathbf{X}_2, \ldots, \mathbf{X}_m$ be a set of $n \times k$ matrices and let the columns of $\mathbf{R}_i$ form an orthonormal basis of $\mathcal{M}(\mathbf{X}_i)$, $i = 1, 2, \ldots, m$. Then **P** is the orthogonal $n \times k$ matrix such that $\overline{\phi}$ is maximal:

$$\overline{\phi} = \frac{1}{m}\sum_{i=1}^{m}\phi_i$$

where $\phi_i = \frac{1}{k}\, tr(\mathbf{P}'\mathbf{R}_i\mathbf{R}_i'\mathbf{P})$. Having this **P**-matrix, put $\Omega = \mathbf{I} - \mathbf{P}\mathbf{P}'$. Here the criterion of choosing Ω is: maximize $\overline{\phi}$. This criterion formalizes expressions such as "Ω as close as possible to $\mathbf{M}_1, \mathbf{M}_2, \ldots, \mathbf{M}_n$

or "$\mathcal{M}(\mathbf{P})$ as close as possible to $\mathcal{M}(\mathbf{X}_1)$, $\mathcal{M}(\mathbf{X}_2), \ldots, \mathcal{M}(\mathbf{X}_m)$". The method of obtaining **P** is known as the method of principal components, as we shall see in the next section. Of course, it is also possible to formulate an analogous criterion in terms of $\psi$. However, we do not know a method of obtaining a matrix **P** from a set of **X**-matrices according to maximization of an average $\psi$.

## 4.3 Principal components

Consider a set of $m$ matrices, $\mathbf{X}_1, \mathbf{X}_2, \ldots, \mathbf{X}_m$, each with rank $k$ and $n$ rows. Let the $n \times k$ matrix $\mathbf{R}_i$ contain an orthonormal basis of $\mathcal{M}(\mathbf{X}_i)$, and let **Z** be the following $n \times mk$ matrix:

$$\mathbf{Z} = [\mathbf{R}_1 \vdots \mathbf{R}_2 \vdots \ldots \vdots \mathbf{R}_m]$$

Then

$$\overline{\phi} = \frac{1}{mk} \sum_{i=1}^{m} tr(\mathbf{P}'\mathbf{R}_i\mathbf{R}_i'\mathbf{P}) = \frac{1}{mk} tr(\mathbf{P}'\mathbf{Z}\mathbf{Z}'\mathbf{P})$$

We wish to find a matrix **P** satisfying $\mathbf{P}'\mathbf{P} = \mathbf{I}_{(k)}$ such that $\overline{\phi}$ is maximal, i.e. $tr(\mathbf{P}'\mathbf{Z}\mathbf{Z}'\mathbf{P})$ must be maximized with respect to **P**, subject to $\mathbf{P}'\mathbf{P} = \mathbf{I}$. This is the well-known principal components problem. The technique for the determination of **P** from **Z** is discussed below. The solution reads:

$$\mathbf{P} = [\mathbf{h}_1 \vdots \mathbf{h}_2 \vdots \ldots \vdots \mathbf{h}_k]$$

where $\mathbf{h}_i$ is the $i$th eigenvector of $\mathbf{Z}\mathbf{Z}'$:

$$\mathbf{Z}\mathbf{Z}' = \mathbf{H}\mathbf{\Lambda}\mathbf{H}' \tag{4.4}$$

with $\mathbf{H}' = \mathbf{H}^{-1}$ and $\lambda_1 \geqslant \lambda_2 \geqslant \ldots \geqslant \lambda_n \geqslant 0$.

The $i$th principal component of **Z** is $\mathbf{h}_i$ (or $c\mathbf{h}_i$, where $c$ is an arbitrary scalar). When $\lambda_i$ is a distinct eigenvalue, then $\mathbf{h}_i$ is unique (see Section 1.A.5). Note that **P** need not be unique. The only relevant aspect of **P** is that $\mathcal{M}(\mathbf{P})$ be a well-defined space. A difficulty arises only when $\lambda_k = \lambda_{k+1}$: then (at least) one of the dimensions of

$\mathcal{M}(\mathbf{P})$ is not completely determined. In our applications we do not come across this case.

We wish to maximize $tr(\mathbf{P}'\mathbf{ZZ}'\mathbf{P})$ with respect to $\mathbf{P}$, subject to $\mathbf{P}'\mathbf{P} = \mathbf{I}_{(k)}$. Observe that $\mathbf{P}$ is not unique if $k > 1$. This can be seen as follows. Let $\mathbf{P}_t = \mathbf{PT}$, where $\mathbf{T}' = \mathbf{T}^{-1}$, then $\mathbf{P}_t'\mathbf{P}_t = \mathbf{I}_{(k)}$ and $tr(\mathbf{P}_t'\mathbf{ZZ}'\mathbf{P}_t) = tr(\mathbf{P}'\mathbf{ZZ}'\mathbf{P})$, so that orthogonal transformations of $\mathbf{P}$ are admissible. The first order conditions for $tr(\mathbf{P}'\mathbf{ZZ}'\mathbf{P})$ to be a maximum follow from differentiating the Lagrange expression:

$$tr(\mathbf{P}'\mathbf{ZZ}'\mathbf{P}) - tr[\mathbf{V}(\mathbf{P}'\mathbf{P} - \mathbf{I})]$$

with respect to $\mathbf{P}$ and $\mathbf{V}$, and putting the derivatives equal to $\mathbf{0}$. This yields $\mathbf{ZZ}'\mathbf{P} = \mathbf{PV}$ and $\mathbf{P}'\mathbf{P} = \mathbf{I}$. Hence $\mathbf{P}'\mathbf{ZZ}'\mathbf{P} = \mathbf{V} = \mathbf{V}' \geqslant \mathbf{0}$, so that $\mathbf{V} = \mathbf{T}\Lambda_{(k)}\mathbf{T}'$ for some $\mathbf{T}$ satisfying $\mathbf{T}' = \mathbf{T}^{-1}$. Then $\mathbf{ZZ}'\mathbf{P} = \mathbf{PV} = \mathbf{PT}\Lambda_{(k)}\mathbf{T}'$, or equivalently, $\mathbf{ZZ}'\mathbf{P}_t = \mathbf{P}_t\Lambda_{(k)}$, where $\mathbf{P}_t = \mathbf{PT}$. It follows that $\mathbf{P}_t$ consists of $k$ eigenvectors and $\Lambda_{(k)}$ contains $k$ eigenvalues. Since we wish to maximize $tr(\mathbf{P}'\mathbf{ZZ}'\mathbf{P}) = tr(\Lambda_{(k)})$, $\Lambda_{(k)}$ should contain the largest $k$ eigenvalues of $\mathbf{ZZ}'$, and then $\mathbf{P}_t$ contains the corresponding eigenvectors of $\mathbf{ZZ}'$.

For a better understanding of principal components, one may think of a linear model with $mk$ observation vectors on the dependent variable, namely the $mk$ columns of $\mathbf{Z}$. The matrix of observations on the explanatory variables, the matrix $\mathbf{P}$, is an artificial matrix, which must be chosen such that the sum of all squared *o.l.s.* residuals is minimal. The matrix of residuals is $\mathbf{Z}^*$:

$$\mathbf{Z}^* = [\mathbf{I} - \mathbf{P}(\mathbf{P}'\mathbf{P})^{-1}\mathbf{P}']\mathbf{Z} = (\mathbf{I} - \mathbf{PP}')\mathbf{Z}$$

and the sum of the squared elements of $\mathbf{Z}^*$ is:

$$\begin{aligned} tr(\mathbf{Z}^*\mathbf{Z}^{*\prime}) &= tr[(\mathbf{I} - \mathbf{PP}')\mathbf{ZZ}'(\mathbf{I} - \mathbf{PP}')] \\ &= tr[(\mathbf{I} - \mathbf{PP}')\mathbf{ZZ}'] \\ &= tr\,(\mathbf{ZZ}') - \mathrm{tr}(\mathbf{P}'\mathbf{ZZ}'\mathbf{P}) \\ &= mk - tr(\mathbf{P}'\mathbf{ZZ}'\mathbf{P}) \end{aligned}$$

$\mathcal{M}(\mathbf{P})$ can be regarded as the best mean regression space with respect to $\mathbf{Z}$, and one may say that $\mathbf{P}$ explains $\mathbf{Z}$ as much as possible. We say that $\mathbf{P}$ explains $\mu$ percent of $\mathbf{Z}$:

$$\mu = 100\, tr(\mathbf{P}'\mathbf{ZZ}'\mathbf{P})\,/mk = 100\,\overline{\phi}$$

and that the $i$th principal component explains $\mu_i$ percent of **Z**:

$$\mu_i = 100\, tr(\mathbf{h}_i'\mathbf{ZZ}'\mathbf{h}_i)\,/mk = 100\,\lambda_i/mk$$

Clearly, $\mu = \mu_1 + \mu_2 + \ldots + \mu_k$ if $\mathbf{P} = [\mathbf{h}_1 \vdots \mathbf{h}_2 \vdots \ldots \vdots \mathbf{h}_k]$. The maximum value of $\mu_i$ is $100/k$, which follows from $0 \leqslant \mathbf{h}_i'\mathbf{M}_j\mathbf{h}_i = \mathbf{h}_i'(\mathbf{I} - \mathbf{R}_j\mathbf{R}_j')\mathbf{h}_i = 1 - \mathbf{h}_i'\mathbf{R}_j\mathbf{R}_j'\mathbf{h}_i$ and hence $\mathbf{h}_i'\mathbf{R}_j\mathbf{R}_j'\mathbf{h}_i \leqslant 1$, so that:

$$\mu_i = 100\, tr(\mathbf{h}_i'\mathbf{ZZ}'\mathbf{h}_i)\,/mk = 100 \sum_{j=1}^{m} \mathbf{h}_i'\mathbf{R}_j\mathbf{R}_j'\mathbf{h}_i/mk \leqslant \frac{100}{k} \quad (4.5)$$

Let **g** be a vector satisfying $\mathbf{g}'\mathbf{g} = 1$. When **g** lies in $\mathcal{M}(\mathbf{R})$, then $\mathbf{Mg} = \mathbf{o}$ and $\mathbf{g}'\mathbf{RR}'\mathbf{g} = \mathbf{g}'(\mathbf{I} - \mathbf{M})\mathbf{g} = 1$. If, on the other hand, **g** is such that $\mathbf{g}'\mathbf{RR}'\mathbf{g} = 1$, then $\mathbf{g}'\mathbf{Mg} = (\mathbf{Mg})'(\mathbf{Mg}) = 0$, so that $\mathbf{Mg} = \mathbf{o}$ and **g** lies in $\mathcal{M}(\mathbf{R})$. It follows that $tr(\mathbf{g}'\mathbf{ZZ}'\mathbf{g}) = m$ if and only if $\mathbf{g}'\mathbf{R}_j\mathbf{R}_j'\mathbf{g} = 1$ for all $j$, which occurs if and only if **g** lies in all regression spaces $\mathcal{M}(\mathbf{R}_j)$. For instance, when all **R**-matrices contain a constant term, then $\mathbf{h}_1$ is a constant term and $\mu_1 = 100/k$.

Above, the principal component problem is solved simultaneously: we found that the $n \times k$ matrix **P** or an orthogonal transformation of it is equal to $[\mathbf{h}_1 \vdots \mathbf{h}_2 \vdots \ldots \vdots \mathbf{h}_k]$, $\mathbf{h}_i$ being the $i$th eigenvector of $\mathbf{ZZ}'$. A successive solution is also possible: first determine the first principal component, then determine the second principal component, and so on, as follows. Let $\mathbf{p}_i$ be the $i$th column of the $n \times l$ matrix $\mathbf{P}_l$ satisfying $\mathbf{P}_l'\mathbf{P}_l = \mathbf{I}_{(l)}$, where $l = 1, 2, \ldots, k, \ldots, n$. The first principal component of **Z** is found from maximization of $tr(\mathbf{P}_1'\mathbf{ZZ}'\mathbf{P}_1) = \mathbf{p}_1'(\sum_{i=1}^{n} \lambda_i\mathbf{h}_i\mathbf{h}_i')\mathbf{p}_1 = \sum_{i=1}^{n} \lambda_i(\mathbf{p}_1'\mathbf{h}_i)^2$ subject to $\mathbf{p}_1'\mathbf{p}_1 = 1$. The solution is $\mathbf{p}_1 = \mathbf{h}_1$. Given $\mathbf{p}_1 = \mathbf{h}_1$, the second principal component of **Z** is found from maximization of $tr(\mathbf{P}_2'\mathbf{ZZ}'\mathbf{P}_2) = \lambda_1 + \sum_{i=2}^{n} \lambda_i(\mathbf{p}_2'\mathbf{h}_i)^2$, which yields $\mathbf{p}_2 = \mathbf{h}_2$. Generally, the $j$th principal component of **Z**, given $\mathbf{p}_1 = \mathbf{h}_1, \ldots, \mathbf{p}_{j-1} = \mathbf{h}_{j-1}$, is found from maximization of $tr(\mathbf{P}_j'\mathbf{ZZ}'\mathbf{P}_j) = \sum_{i=1}^{j-1} \lambda_i + \sum_{i=j}^{n} \lambda_i(\mathbf{p}_j'\mathbf{h}_i)^2$, yielding $\mathbf{p}_j = \mathbf{h}_j$. Denoting by $\mathbf{Z}_j^*$ the residual matrix of **Z** after explanation by the first $j$ principal components:

$$\mathbf{Z} = \mathbf{P}_j\mathbf{P}_j'\mathbf{Z} + \mathbf{Z}_j^*$$

we have $\mathbf{I} - \mathbf{P}_j\mathbf{P}_j' = \mathbf{I} - \mathbf{h}_1\mathbf{h}_1' - \ldots - \mathbf{h}_j\mathbf{h}_j' = \mathbf{h}_{j+1}\mathbf{h}_{j+1}' + \ldots + \mathbf{h}_n\mathbf{h}_n'$ and:

$$\begin{aligned}\mathbf{Z}_j^*\mathbf{Z}_j^{*\prime} &= (\mathbf{I} - \mathbf{P}_j\mathbf{P}_j')\mathbf{Z}\mathbf{Z}'(\mathbf{I} - \mathbf{P}_j\mathbf{P}_j') \\ &= (\mathbf{h}_{j+1}\mathbf{h}_{j+1}' + \ldots + \mathbf{h}_n\mathbf{h}_n')(\sum_{i=1}^{n}\lambda_i\mathbf{h}_i\mathbf{h}_i')(\mathbf{h}_{j+1}\mathbf{h}_{j+1}' + \ldots + \mathbf{h}_n\mathbf{h}_n') \\ &= \sum_{i=j+1}^{n}\lambda_i\mathbf{h}_i\mathbf{h}_i'\end{aligned}$$

so that the $(j+1)$-th principal component of **Z** is equal to the first principal component of $\mathbf{Z}_j^*$. Hence, the first $k$ principal components of **Z** are successively found as the first principal components of **Z**, $\mathbf{Z}_1^*, \ldots, \mathbf{Z}_{k-1}^*$, respectively.

## 4.4 An empirical P-matrix

We wish to establish a **P**-matrix, to be applied in economic time series analysis. To this end we collected time series data for the construction of a **Z**-matrix. In accordance with the previous section, the first $k$ principal components of **Z** constitute **P**. Below we present the first six principal components of an empirical **Z**-matrix. The section is concluded by some remarks.

We collected sixty vectors, each consisting of $n = 15$ subsequent annual observations on economic variables. Arbitrarily splitting up this collection into thirty pairs of vectors, and adding a constant term to each pair, we got $m = 30$ **X**-matrices with $k = 3$ column vectors. The data include both stock and flow variables, deflated and undeflated, price indices, and also logarithmic series. First order and higher order differences are not included. Every vector covers a 15-year period between 1920 and 1969, and contains either American, English or Dutch data.

From these data a $15 \times 90$ matrix **Z** can be constructed. In Table 4.1, the first six principal components $\mathbf{h}_i$ of **Z** are presented, together with $\mu_i$, their percentage explanation of **Z**. The vectors are graphically presented in Figure 4.1.

*Table 4.1* **The first six principal components of Z.**

| | $\mathbf{h}_1$ | $\mathbf{h}_2$ | $\mathbf{h}_3$ | $\mathbf{h}_4$ | $\mathbf{h}_5$ | $\mathbf{h}_6$ |
|---|---|---|---|---|---|---|
| 1 | 0.258 | 0.512 | 0.145 | 0.407 | 0.212 | 0.006 |
| 2 | 0.258 | 0.475 | 0.094 | 0.148 | 0.121 | 0.040 |
| 3 | 0.258 | 0.359 | 0.087 | - 0.049 | - 0.062 | - 0.192 |
| 4 | 0.258 | 0.197 | - 0.170 | - 0.234 | - 0.493 | - 0.249 |
| 5 | 0.258 | 0.098 | - 0.216 | - 0.440 | - 0.211 | 0.190 |
| 6 | 0.258 | 0.018 | - 0.236 | - 0.328 | 0.132 | 0.270 |
| 7 | 0.258 | - 0.097 | - 0.240 | - 0.137 | 0.387 | 0.307 |
| 8 | 0.258 | - 0.178 | - 0.275 | 0.083 | 0.170 | - 0.057 |
| 9 | 0.258 | - 0.245 | - 0.327 | 0.325 | 0.116 | - 0.245 |
| 10 | 0.258 | - 0.279 | - 0.202 | 0.309 | - 0.239 | - 0.270 |
| 11 | 0.258 | - 0.226 | 0.041 | 0.293 | - 0.184 | 0.271 |
| 12 | 0.258 | - 0.178 | 0.312 | 0.110 | - 0.015 | 0.499 |
| 13 | 0.258 | - 0.116 | 0.414 | - 0.025 | - 0.380 | 0.068 |
| 14 | 0.258 | - 0.147 | 0.404 | - 0.129 | - 0.005 | - 0.197 |
| 15 | 0.258 | - 0.194 | 0.342 | - 0.336 | 0.452 | - 0.441 |
| $\mu_i$ | 33.3 | 27.1 | 17.5 | 8.5 | 4.1 | 3.6 |

The empirical 15 × 3 matrix $\mathbf{P} = [\mathbf{h}_1 \vdots \mathbf{h}_2 \vdots \mathbf{h}_3]$ explains

$$\mu = \mu_1 + \mu_2 + \mu_3 = 33.3 + 27.1 + 17.5 = 77.9$$

percent of **Z**, and hence the maximum value of $\overline{\phi}$ is 0.779 for this empirical **Z**.

The first remark concerns the order of **P**. Our approach involves that **P** has the same order as **X**, provided that the columns of **X** are linearly independent. As our range of application we consider $n \times k$ **X**-matrices with $10 \leqslant n \leqslant 20$ and $2 \leqslant k \leqslant 4$, implying 33 **P**-matrices, one for each combination of $n$ and $k$. It would be much more attractive if one **P**-matrix would do; for instance, one $p \times (p\text{-}r)$ matrix **P** with $p = 15$ for all **X**-matrices under consideration. Then **w** must have 15 elements, regardless of $n$. This can be achieved by replacing (4.1) by $\mathbf{w} = \mathbf{K}(\mathbf{K}'\mathbf{J}'\mathbf{M}\mathbf{J}\mathbf{K})^{-1/2}\mathbf{K}'\mathbf{J}'\mathbf{M}\mathbf{y}$, where **J** is an $n \times 15$ matrix. However, $r = rank(\Omega)$ cannot exceed $n\text{-}k$, so that $r \leqslant 6$ (for $n = 10$ and

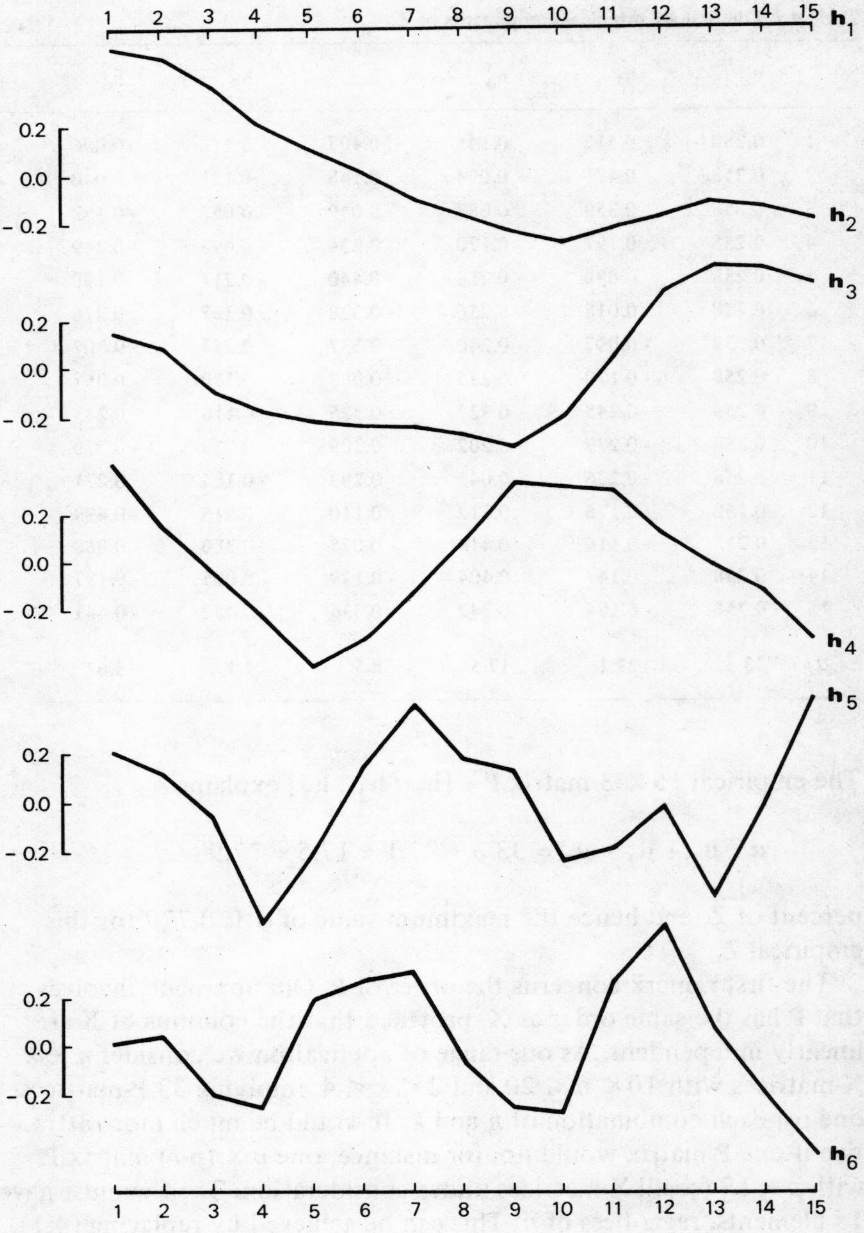


*Figure 4.1* **Diagram of the first six principal components of Z.**

$k = 4$), which implies that $r$ is usually much smaller than $n$-$k$. In Section 5.4 we make some experiments with the matrix **J**. The powers in that section reveal that it is a disastrous to take $r < n$-$k$, and that $p \neq n$ also leads to a loss of power.

The second remark concerns the validity of a **P**-matrix. We wish to have a **P**-matrix, which is to be used in models with **R**-matrices, which are not included in our empirical **Z**. Considering that another empirical **Z** would yield a **P**-matrix which is not exactly equal to our empirical **P**, we feel that small changes in **P** are admissible. After a streamlining procedure in the next section, we adopt a **P**-matrix, which is close to our empirical **P**-matrix (the streamlined matrix has great advantages). It explains as much as 76.3 percent of **Z**, while the principal components **P** explains 77.9 percent of **Z**, which is the maximum attainable. Incidentally, in applications of a **P**-matrix to empirically specified models, we may meet **X**-matrices such that all $\phi$-values are very small. It is impossible to make probability statements about such events. Fortunately, the mean of the $\phi$-values in our applications is 0.75 (see Table 5.1, the column under the heading *hyp/z*; tne same value arises as the central value of $\overline{\phi}$ in Section 4.6). In our opinion, the small differences between the $\overline{\phi}$-values, at the level of 0.75, support the statement that the streamlined **P** may be regarded as typical of economic time series. The third remark concerns $n \times k$ **P**-matrices with $n = 15$ and $k \neq 3$ in the first place, and $n \neq 15$ and $k = 3$ in the second place. In preliminary investigations we regrouped the collected data such that we got a number of $15 \times 4$ **X**-matrices, or $15 \times 5$ **X**-matrices, or $12 \times 3$ **X**-matrices, or $10 \times 3$ **X**-matrices. The diagrams of the relevant principal components of the corresponding **Z**-matrices showed so much resemblance to Figure 4.1 that it seems worthwhile to seek a formula which generates the $i$th column of a streamlined $n \times k$ **P**-matrix for arbitrary $n$ and $k$. If we find such a formula, then the calculation of **w** can be facilitated by a computer subroutine, which generates **P** for given $n$ and $k$. The generalization with respect to $n$ and $k$ of the streamlined **P**-matrix in the next section is the subject of Section 4.6.

## 4.5 Streamlining of P

In Figure 4.1, one immediately recognizes waves with various frequencies, and the ordering is striking: $\mathbf{h}_1$ is a degenerate wave, $\mathbf{h}_2$ can be characterized by "down", $\mathbf{h}_3$ by "down-up", $\mathbf{h}_4$ by "down-up-down", and so on. Vector $\mathbf{h}_1$ is a perfect constant term, which explains $\mu_1$ = 33.3 percent of **Z** (see Table 4.1). This is a consequence of the inclusion of a constant term in all **X**-matrices, see the discussion on page 71. Vector $\mathbf{h}_1$ needs, of course, no streamlining.

Vector $\mathbf{h}_2$ deviates from a regular trend in a rather peculiar way. It seems to account partially for a cyclical component. There is no reason to assume that this divergence is typical of economic time series. As a first step, we streamline the picture of $\mathbf{h}_2$ to a straight line, say $\mathbf{s}_2$, which vector can be obtained as follows. Let **a** be the 15-element vector whose *i*th element is equal to *i*. Let $\mathbf{a}^* = (\mathbf{I} - \mathbf{h}_1\mathbf{h}_1')$ **a**, so that $\mathbf{h}_1'\mathbf{a}^* = 0$, then $\mathbf{s}_2 = -(\mathbf{a}^{*\prime}\mathbf{a}^*)^{-½}\mathbf{a}^*$ (the minus sign is added in order that the first element of $\mathbf{s}_2$ be positive, in accordance with our convention with respect to eigenvectors). However, when we replace $\mathbf{h}_2$ by $\mathbf{s}_2$, then $\mathbf{h}_3, \mathbf{h}_4, \ldots$ cannot be maintained, since $\mathbf{s}_2'\mathbf{h}_i \neq 0$ for $i = 3, 4, \ldots$, generally. If it is given that the first and the second column of **P** are $\mathbf{h}_1$ and $\mathbf{s}_2$, respectively, and **P** is supposed to explain as much as possible of **Z**, then the third column of **P** must be the first principal component of $\mathbf{Z}_2^{**}$, the residual matrix of **Z** after explanation by $\mathbf{h}_1$ and $\mathbf{s}_2$,

$$\mathbf{Z}_2^{**} = (\mathbf{I} - \mathbf{h}_1\mathbf{h}_1' - \mathbf{s}_2\mathbf{s}_2')\mathbf{Z}$$

Figure 4.2 displays $\mathbf{h}_1$, $\mathbf{s}_2$, and the first four principal components of $\mathbf{Z}_2^{**}$. The percentage contributions of these vectors to the explanation of **Z** are, respectively:

33.3 24.2 20.0 8.6 4.1 3.8

Comparing these percentages with those in the bottom row of Table 4.1, we see that $\mathbf{s}_2$ explains 3 percent less than $\mathbf{h}_2$, this loss being almost completely regained (compared with $\mathbf{h}_3$) by the first principal component of $\mathbf{Z}_2^{**}$. The matrix **P** consisting of $\mathbf{h}_1$, $\mathbf{s}_2$, and the first principal component of $\mathbf{Z}_2^{**}$ explains 77.5 percent of **Z**, a loss of only 0.4 percent compared with $\mathbf{P} = [\mathbf{h}_1 : \mathbf{h}_2 : \mathbf{h}_3]$. Comparing $\mathbf{h}_3$,

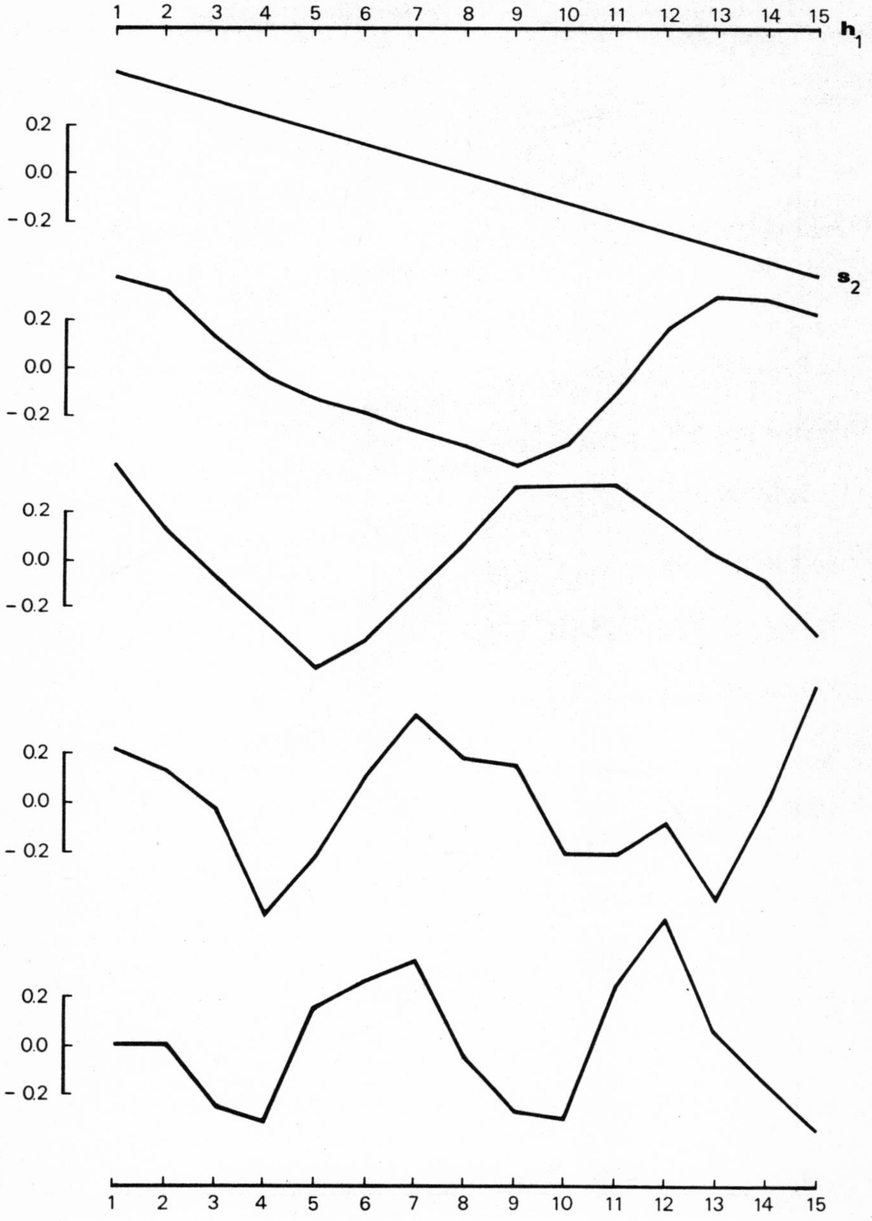


*Figure 4.2* $\mathbf{h}_1$, $\mathbf{s}_2$, and the first four principal components of $\mathbf{Z}_2^{**}$.

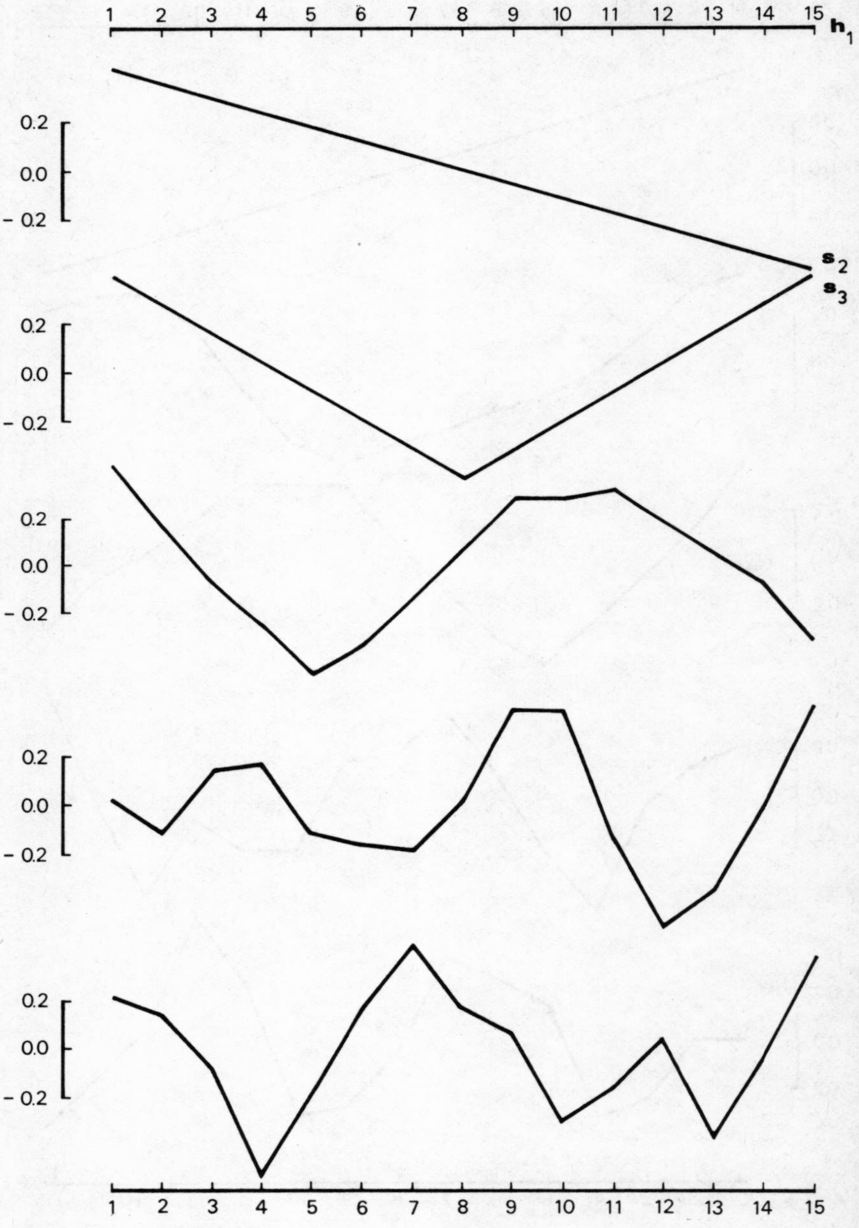


*Figure 4.3* $\mathbf{h}_1$, $\mathbf{s}_2$ and $\mathbf{s}_3$, and the first three principal components of $\mathbf{Z}_3^{**}$.

$\mathbf{h}_4$, $\mathbf{h}_5$, and $\mathbf{h}_6$ in Figure 4.1 with their respective counterparts in Figure 4.2, we see that $\mathbf{h}_4$, $\mathbf{h}_5$ and $\mathbf{h}_6$ are very stable, just like their percentage contribution to the explanation of **Z**. The picture of $\mathbf{h}_3$ has been changed, in favour of a more regular down-up pattern. This supports the next streamlining. Vector $\mathbf{s}_3$ in Figure 4.3 is obtained as follows. Take $\mathbf{a}' = [7\ 6 \ldots 1\ 0\ 1 \ldots 6\ 7]$, $\mathbf{a}^* = (\mathbf{I} - \mathbf{h}_1\mathbf{h}_1' - \mathbf{s}_2\mathbf{s}_2')\mathbf{a}$, $\mathbf{s}_3 = (\mathbf{a}^{*\prime}\mathbf{a}^*)^{-1/2}\mathbf{a}^*$. Compared with Figure 4.1, we have replaced $\mathbf{h}_4$, $\mathbf{h}_5$, and $\mathbf{h}_6$ by the first three principal components of $\mathbf{Z}_3^{**}$:

$$\mathbf{Z}_3^{**} = (\mathbf{I} - \mathbf{h}_1\mathbf{h}_1' - \mathbf{s}_2\mathbf{s}_2' - \mathbf{s}_3\mathbf{s}_3')\mathbf{Z}$$

The percentage contributions to the explanation of **Z** now are, respectively:

33.3 24.2 18.1 8.6 5.0 4.1

Comparing Figures 4.1 and 4.3, we conclude that $\mathbf{h}_2$ and $\mathbf{h}_3$ can be replaced by $\mathbf{s}_2$ and $\mathbf{s}_3$ at low cost (in the sense of explanatory contribution) and that $\mathbf{h}_4$, $\mathbf{h}_5$, and $\mathbf{h}_6$ are very stable, the patterns of $\mathbf{h}_5$ and $\mathbf{h}_6$ being interchanged. Note that $\mathbf{P} = [\mathbf{h}_1 : \mathbf{s}_2 : \mathbf{s}_3]$ explains 75.6 percent of **Z**. We terminate this type of streamlining. In the first place because of the difficulties arising when constructing $\mathbf{s}_4$, $\mathbf{s}_5$, and so on. Where should the two turning points in a streamlined down-up-down movement be located when $n = 15$? And where the three turning points in the next down-up-down-up movement? And how to generalize such movements with respect to $n$? In the second place, smooth turning points are probably more realistic in economics than sharp turning points. Therefore we look for another type of streamlining. The above streamlining procedure is important in that it provides us with some experience: $\mathbf{h}_2$ and $\mathbf{h}_3$, together explaining 44.6 percent of **Z**, can be replaced by $\mathbf{s}_2$ and $\mathbf{s}_3$, together explaining 42.3 percent, causing a loss of 2.3 percent, while $\mathbf{h}_4$, $\mathbf{h}_5$, and $\mathbf{h}_6$ are practically unaffected. We think that the loss of 2.3 percent is to be regarded as a small loss, both in view of the level of 44.6 percent and in view of the fact that the superiority (in the sense of percentage explanation of **Z**) of $\mathbf{s}_3$ (18.1 percent) relative to the first principal component of $\mathbf{Z}_3^{**}$ (8.6 percent) remains impressive.

We considered sets of polynomials. The *Chebyshev polynomials* (see Abramovitz and Stegun 1965, figure 22.6) seem appropriate.

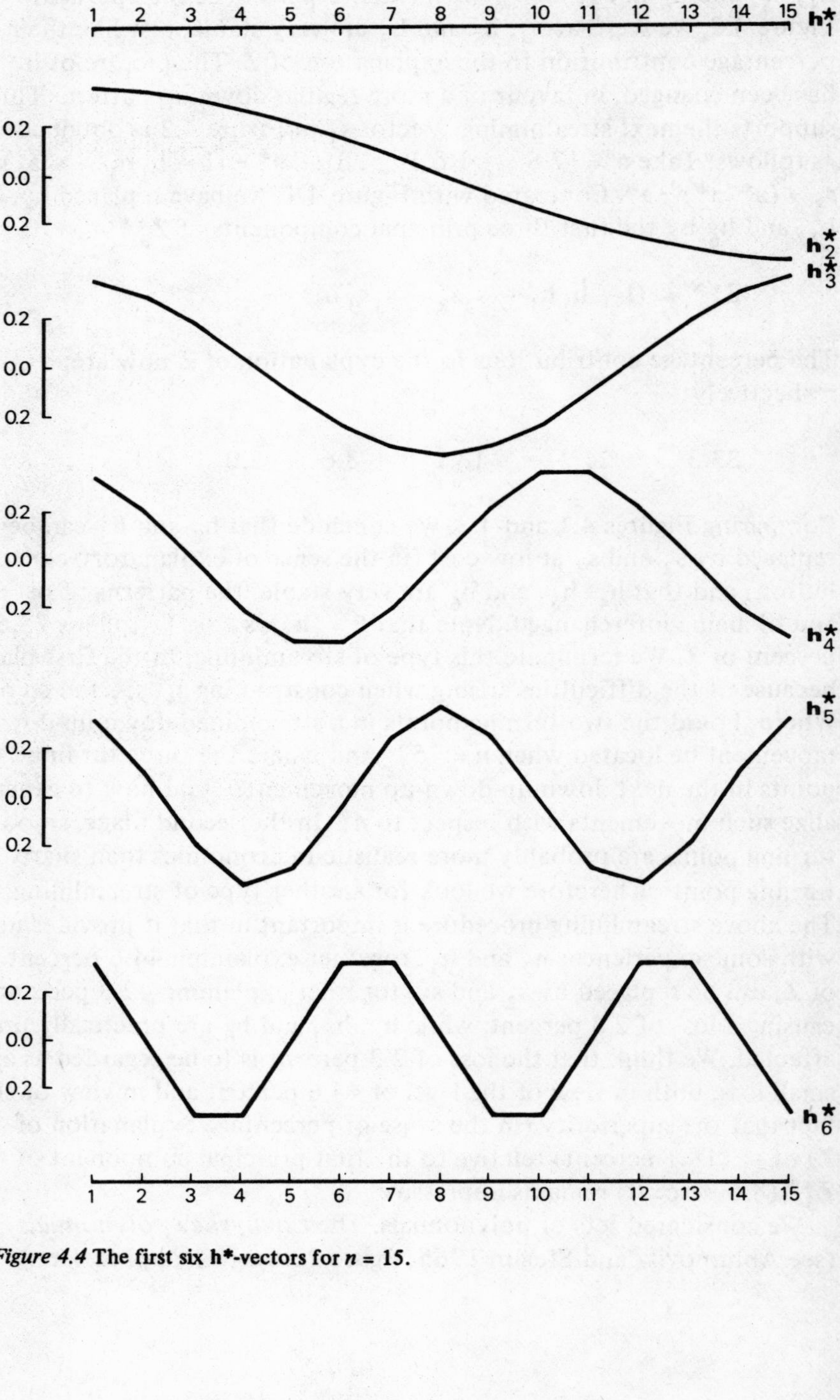


*Figure 4.4* **The first six h*-vectors for *n* = 15.**

Such a polynomial of degree $t$ is defined as:

$$C_t(x) = \cos\left\{t \arccos(x)\right\} \qquad |x| \leqslant 1$$

Hence, $C_0(x) = 1$, $C_1(x) = x$, $C_2(x) = 2x^2 - 1$, $C_3(x) = 4x^3 - 3x$, $C_4(x) = 8x^4 - 8x^2 + 1$, and so on, where $C_t(x)$ for $t > 1$ is most easily found from the recursion formula:

$$C_{t+1}(x) = 2xC_t(x) - C_{t-1}(x)$$

which follows from the definition of $C_t(x)$ and the trigonometric formula:

$$2\cos(\theta)\cos(t\theta) = \cos\left\{(t+1)\theta\right\} + \cos\left\{(t-1)\theta\right\}$$

by taking $\cos(\theta) = x$, so that $\theta = \arccos(x)$, and $C_t(x) = \cos(t\theta)$. With $0 \leqslant \theta \leqslant \pi$ so as to obtain a one-to-one correspondence between $x$ and $\theta$, it is seen that $C_t(x) = 0$ for $\theta = \theta_j = \pi(j + \tfrac{1}{2})/t$ with $j = 0, 1, 2, \ldots, t$-1. Hence, $C_t(x) = 0$ for $x = x_j = \cos\left\{\pi(j + \tfrac{1}{2})/t\right\}$ with $j = 0, 1, 2, \ldots, t$-1. It is known (see Hildebrand 1956, p. 390) that the Chebyshev polynomials are orthogonal in the following sense:

$$\begin{aligned}\sum_{j=0}^{n-1} C_t(x_j)C_s(x_j) &= 0 && \text{if } t \neq s \\ &= n && \text{if } t = s = 0 \qquad (4.6)\\ &= n/2 && \text{if } t = s \neq 0\end{aligned}$$

where $t = 0, 1, 2, \ldots, n$-1 and $s = 0, 1, 2, \ldots, n$-1, and where $x_j$, $j = 0, 1, 2, \ldots, n$-1, are the values of $x$ such that $C_n(x) = 0$. Thus, the $n$-element vector whose $j$th element is $C_t(x_{j-1})$ has length $n^{1/2}$ if $t = 0$ and length $(n/2)^{1/2}$ if $t = 1, 2, \ldots, n$-1. Since $x_{j-1} = \cos\left\{\pi(j - \tfrac{1}{2})/n\right\}$, we have $C_{i-1}(x_{j-1}) = \cos\left\{\pi(i-1)(j - \tfrac{1}{2})/n\right\}$.

We define the $j$th element of the $n$-element vector $\mathbf{h}_i^*$, this vector being the $i$th column of the $n \times n$ matrix $\mathbf{H}^*$, as:

$$h_i^*(j) = c \cos\left\{ \pi(i-1)(j-\tfrac{1}{2})/n \right\} \qquad c = n^{-\frac{1}{2}} \quad \text{if } i = 1$$
$$c = \left(\frac{n}{2}\right)^{-\frac{1}{2}} \quad \text{if } i \neq 1 \tag{4.7}$$

According to the orthogonality theorem (4.6) we have $\mathbf{H}^{*\prime} = \mathbf{H}^{*-1}$. Figure 4.4 shows $\mathbf{h}_1^*$ through $\mathbf{h}_6^*$ for $n = 15$. These vectors explain, respectively,

33.3 24.3 18.7 8.2 3.8 4.2

percent of $\mathbf{Z}$. The matrix $[\mathbf{h}_1^* : \mathbf{h}_2^* : \mathbf{h}_3^*]$ explains 76.3 percent of $\mathbf{Z}$, which is only 1.6 percent below the maximum of 77.9 percent, attained by $[\mathbf{h}_1 : \mathbf{h}_2 : \mathbf{h}_3]$. We adopt $\mathbf{h}_i^*$ as the idealization of $\mathbf{h}_i$. These vectors happen to be eigenvectors of $\mathbf{A}_d$, see (2.24). On the basis of remarks made by Hannan (1960), the vectors $\mathbf{h}_1^*$, $\mathbf{h}_2^*$, and $\mathbf{h}_3^*$ were adopted and applied to tests against positive autocorrelation using (4.1) by Abrahamse and Louter (1971).

## 4.6 Generalization for *n* and *k*

Using the same empirical data as in Section 4.4, we attempt to justify the generalization from the $15 \times 3$ matrix $\mathbf{P} = [\mathbf{h}_1 : \mathbf{h}_2 : \mathbf{h}_3]$ to the $n \times k$ matrix $\mathbf{P} = [\mathbf{h}_1^* : \mathbf{h}_2^* : \ldots : \mathbf{h}_k^*]$, for $n$ and $k$ in the neighbourhood of 15 and 3, respectively.

*Generalization for k*

Let $\mathbf{Z}(m; n, k)$ denote an $n \times mk$ matrix consisting of $m$ $\mathbf{R}$-matrices, each of order $n \times k$. In the preceding sections we investigated $\mathbf{Z}(30; 15, 2+1)$, where $2 + 1$ indicates that the original $\mathbf{X}$-matrices contain two time series columns and a constant term column. Rearrangement of our sixty data vectors yielded $\mathbf{Z}(20; 15, 3)$ and $\mathbf{Z}(20; 15, 3+1)$, the only difference being a constant term column in all $\mathbf{X}$-matrices. Also we got $\mathbf{Z}(15; 15, 4)$ and $\mathbf{Z}(15; 15, 4+1)$, with the same difference, and $\mathbf{Z}(12; 15, 5)$. Let $\mu_i^*$ denote the percentage explanation of $\mathbf{Z}(m; n, k)$ by $\mathbf{h}_i^*$, like $\mu_i$ denotes the percentage explanation of that matrix by its $i$th principal component. The values of $\mu_i$ and $\mu_i^*$ for $i = 1, 2, \ldots, 6$ are presented in Table 4.2, where the bottom row contains the

*Table 4.2* **Percentage explanation of Z $(m; n, k)$ by its $i$th principal component and by $h_i^*$.**

| | Z(30;15,2+1) | | Z(20;15,3+1) | | Z(15;15,4+1) | | Z(20;15,3) | | Z(15;15,4) | | Z(12;15,5) | |
|---|---|---|---|---|---|---|---|---|---|---|---|---|
| $i$ | $\mu_i$ | $\mu_i^*$ | $\mu_i$ | $\mu_i^*$ | $\mu_i$ | $\mu_i^*$ | $\mu_i$ | $\mu_i^*$ | $\mu_i$ | $\mu_i^*$ | $\mu_i$ | $\mu_i^*$ |
| 1 | 33.3 | 33.3 | 25.0 | 25.0 | 20.0 | 20.0 | 33.1 | 33.1 | 24.9 | 24.8 | 20.0 | 20.0 |
| 2 | 27.1 | 24.3 | 22.0 | 21.6 | 18.8 | 18.4 | 25.5 | 25.2 | 21.4 | 21.1 | 18.5 | 18.3 |
| 3 | 17.5 | 18.7 | 18.1 | 17.2 | 17.2 | 16.0 | 17.6 | 16.1 | 20.7 | 18.7 | 17.1 | 15.8 |
| 4 | 8.5 | 8.2 | 11.8 | 11.4 | 13.4 | 13.2 | 8.4 | 8.4 | 13.4 | 13.3 | 12.8 | 12.9 |
| 5 | 4.1 | 3.8 | 7.2 | 6.3 | 8.8 | 7.9 | 5.0 | 3.8 | 5.2 | 3.4 | 9.2 | 7.0 |
| 6 | 3.6 | 4.2 | 5.9 | 6.4 | 6.1 | 6.5 | 4.0 | 5.0 | 4.1 | 5.8 | 7.0 | 8.1 |
| $\sum_1^k$ | 77.9 | 76.4 | 76.9 | 75.2 | 78.2 | 75.6 | 76.3 | 74.3 | 80.5 | 77.9 | 77.5 | 74.0 |

values of $\sum_{i=1}^{k} \mu_i$ and $\sum_{i=1}^{k} \mu_i^*$, respectively, for all **Z**-matrices mentioned above.

The subcolumns below $\mu_i$ and $\mu_i^*$ in Table 4.2 are as pairs very much alike. The conclusion is that $\mathbf{P} = [\mathbf{h}_1 \vdots \mathbf{h}_2 \vdots \ldots \vdots \mathbf{h}_k]$ can very well be replaced by $\mathbf{P} = [\mathbf{h}_1^* \vdots \mathbf{h}_2^* \vdots \ldots \vdots \mathbf{h}_k^*]$ for $n = 15$, at least for $k = 3, 4, 5$. The latter **P**-matrix roughly explains 75 percent of **Z** in all cases.

*Generalization for n*

We examine four types of submatrices of the **X**-matrices underlying **Z**(30; 15, 2+1): delete (*a*) the first three, (*b*) the first five, (*c*) the last three, and (*d*) the last five rows of each **X**. For each type we constructed a matrix **Z**(30; $n$, 2+1), denoted by $\mathbf{Z}_a$, $\mathbf{Z}_b$, $\mathbf{Z}_c$, and $\mathbf{Z}_d$, with $n$ = 12, 10, 12, and 10, respectively. The conclusion from Table 4.3 is similar to that from Table 4.2: $\mathbf{P} = [\mathbf{h}_1 \vdots \mathbf{h}_2 \vdots \mathbf{h}_3]$ can very well be replaced by $\mathbf{P} = [\mathbf{h}_1^* \vdots \mathbf{h}_2^* \vdots \mathbf{h}_3^*]$, at least for $n$ = 10, 12, 15. Both the variation among the values in the bottom row of Table 4.3 and the average value are somewhat larger, compared with Table 4.2. Still, 75 percent seems to be a rather general average explanation level.

*Table 4.3* **Percentage explanation of $\mathbf{Z}_j$ by its *i*th principal component and by $\mathbf{h}_i^*$.**

| | $\mathbf{Z}_a$ | | $\mathbf{Z}_b$ | | $\mathbf{Z}_c$ | | $\mathbf{Z}_d$ | |
|---|---|---|---|---|---|---|---|---|
| $i$ | $\mu_i$ | $\mu_i^*$ | $\mu_i$ | $\mu_i^*$ | $\mu_i$ | $\mu_i^*$ | $\mu_i$ | $\mu_i^*$ |
| 1 | 33.3 | 33.3 | 33.3 | 33.3 | 33.3 | 33.3 | 33.3 | 33.3 |
| 2 | 23.7 | 23.4 | 24.8 | 24.5 | 30.3 | 29.1 | 30.7 | 30.4 |
| 3 | 20.4 | 18.5 | 18.6 | 15.9 | 17.3 | 16.7 | 17.4 | 17.2 |
| 4 | 8.6 | 8.6 | 10.3 | 12.6 | 8.5 | 6.9 | 6.8 | 6.6 |
| 5 | 5.6 | 6.3 | 5.3 | 3.7 | 4.6 | 6.4 | 4.8 | 4.2 |
| 6 | 2.9 | 2.7 | 3.1 | 4.0 | 2.1 | 2.3 | 2.8 | 1.5 |
| $\sum_1^3$ | 77.4 | 75.3 | 76.7 | 73.7 | 80.6 | 79.1 | 81.4 | 80.9 |

## 4.7 An empirical hypothesis and a selection device

The results of the preceding sections suggest the following hypothesis:

$$\mathbf{P} = [\mathbf{h}_1^* \vdots \mathbf{h}_2^* \vdots \ldots \vdots \mathbf{h}_k^*] \tag{4.8}$$

For this **P**, $\mathbf{I} - \mathbf{PP}' = \mathbf{KK}' = \Omega$ is idempotent and, on average, it is as close as possible to **M** (in a least-squares sense) if the only thing we know about **X** is that it contains annual economic time series data. Note that **K** is most easily found as $[\mathbf{h}_{k+1}^* \vdots \mathbf{h}_{k+2}^* \vdots \ldots \vdots \mathbf{h}_n^*]$. We identify the vector **w**, see (4.1), using this specification of **P**, by *hyp*. In applications to *test(T)*, see (2.1), we shall speak of: *test(T)* using *hyp*, see also Section 5.1.

From our testing results in Chapter 5 we learn that the power of *test(Q)* using *hyp* is usually very satisfactory. However, for some **X**-matrices the power is disappointingly small. This experience brings us to the following concept.

Suppose that we are given a class of **P**-matrices containing the **P** in (4.8). When a test must be performed, then we do not only know that **X** contains annual economic time series data, since **X** is completely specified. This knowledge can be used to choose one **P** from the class. For each **P**-matrix, $\mathbf{P} = [\mathbf{p}_1 \vdots \mathbf{p}_2 \vdots \cdots \vdots \mathbf{p}_k]$, we may calculate the value of $\phi$, see (4.3):

$$k\phi = tr(\mathbf{P}'\mathbf{R}\mathbf{R}'\mathbf{P}) = tr[\mathbf{P}'\mathbf{X}(\mathbf{X}'\mathbf{X})^{-1}\mathbf{X}'\mathbf{P}] = \sum_{i=1}^{k} \mathbf{p}_i'\mathbf{X}(\mathbf{X}'\mathbf{X})^{-1}\mathbf{X}'\mathbf{p}_i$$

The best **P** with respect to the given **X**-matrix is the one yielding the greatest $\phi$-value. When a large number of **X**-matrices is given, we expect that in a majority of cases the **P** in (4.8) emerges as the best **P**. But in some instances other **P**-matrices may better fit. The problem is, how to define the class of **P**-matrices. We propose to consider all $n \times k$ **P**-matrices consisting of **h***-vectors. Then **P** in accordance with:

$$\mathbf{P} = [\mathbf{h}^*_{i1} \vdots \mathbf{h}^*_{i2} \vdots \ldots \vdots \mathbf{h}^*_{ik}], \text{ with } i1, i2, \ldots, ik \text{ such that:} \tag{4.9}$$

$$\phi^*_{i1} \geqslant \phi^*_{i2} \geqslant \ldots \geqslant \phi^*_{ik} \geqslant \ldots \geqslant \phi^*_{in}, \text{ where:}$$

$$\phi^*_j = \mathbf{h}^{*\prime}_j\mathbf{X}(\mathbf{X}'\mathbf{X})^{-1}\mathbf{X}'\mathbf{h}^*_j/k \text{ for } j = 1, 2, \ldots, n$$

is the best P with respect to the given X, since $\phi = \sum_{j=i1}^{ik} \phi^*_j$.

The device to select this **P** from the proposed class thus requires calculation of the diagonal elements of $\mathbf{H}^{*\prime}\mathbf{X}(\mathbf{X}'\mathbf{X})^{-1}\mathbf{X}'\mathbf{H}^*$, and then **P** consists of the $k$ **h***-vectors corresponding to the $k$ greatest diagonal elements. Computationally, the selection device is simple, partly owing to the composition of the proposed class. The choice of this class is rather arbitrary. We may argue that **Z** in Section 4.4 could be explained up to 77.9 percent by $\mathbf{h}_1$, $\mathbf{h}_2$ and $\mathbf{h}_3$ together, so that 22.1 percent remains unexplained, and that $\mathbf{h}_4$, $\mathbf{h}_5$, and $\mathbf{h}_6$ together explain 16.2 percent of **Z**. Less than 6 percent can be explained by $\mathbf{h}_7$, $\mathbf{h}_8, \ldots$, and $\mathbf{h}_{15}$ together. The last nine principal components probably just fit the residual matrix without showing recognizable structures. But $\mathbf{h}_4$, $\mathbf{h}_5$, and $\mathbf{h}_6$ could well be streamlined to $\mathbf{h}^*_4$, $\mathbf{h}^*_5$, and $\mathbf{h}^*_6$, while the value of $\mu^*_4$ and, to a lesser extent, the values of $\mu^*_5$ and $\mu^*_6$, are great enough that these **h***-vectors may become important in the explanation of individual **X**-matrices. The higher frequency vectors have not been introduced as streamlined principal components; they simply span the residual space in a well-defined way. In tests applied to quarterly or monthly data rather than annual data, **P**-matrices, containing higher frequency vectors, probably better fit. We do not consider such data in this study.

## 4.A. Appendix

In this appendix some computational aspects of **w** in (4.1) with $\Omega$ idempotent are considered. Below we prove:

(4.A.1) $(\mathbf{K}'\mathbf{M}\mathbf{K})^{-1/2}$ exists if and only if $\mathbf{P}'\mathbf{R}$ is nonsingular.

(4.A.2) $tr(\mathbf{K}'\mathbf{M}\mathbf{K})^{1/2} = n - 2k + tr(\mathbf{P}'\mathbf{R}\mathbf{R}'\mathbf{P})^{1/2} = n - 2k + \sum_{i=1}^{k} d_i^{1/2}$

(4.A.3) $(\mathbf{K}'\mathbf{M}\mathbf{K})^{-1/2} = \mathbf{I} + (\mathbf{K}'\mathbf{M}\mathbf{P}\mathbf{L}\mathbf{D}^{-1/2})(\mathbf{D} + \mathbf{D}^{1/2})^{-1}(\mathbf{K}'\mathbf{M}\mathbf{P}\mathbf{L}\mathbf{D}^{-1/2})'$

(4.A.4) $\mathbf{B}' = \mathbf{I} - \mathbf{U}\mathbf{D}^{-1}\mathbf{U}' - \mathbf{V}(\mathbf{I} + \mathbf{D}^{1/2})^{-1}\mathbf{V}' - \mathbf{U}\mathbf{D}^{-1/2}\mathbf{V}'$

(4.A.5) $\mathbf{w} = \hat{\mathbf{u}} - \mathbf{P}\mathbf{L}(\mathbf{I} + \mathbf{D}^{1/2})^{-1}\mathbf{L}'\mathbf{P}'\hat{\mathbf{u}} - \mathbf{X}(\mathbf{X}'\mathbf{X})^{-1}\mathbf{X}'\mathbf{P}\mathbf{L}\mathbf{D}^{-1/2}$
$(\mathbf{I} + \mathbf{D}^{1/2})^{-1}\mathbf{L}'\mathbf{P}'\hat{\mathbf{u}}$

where **D**, **L**, **U**, and **V** are defined by:

$\mathbf{P}'\mathbf{R}\mathbf{R}'\mathbf{P} = \mathbf{L}\mathbf{D}\mathbf{L}'$, with $\mathbf{L}' = \mathbf{L}^{-1}$, **D** diagonal.
$\mathbf{V} = \mathbf{M}\mathbf{P}\mathbf{L}$
$\mathbf{U} = \mathbf{P}\mathbf{L} - \mathbf{V}$

A Fortran computer program for the calculation of **w** from **X** and **y**, with **P** determined by the program in accordance with (4.9), is given in Louter and Dubbelman (1973).

*Proof of (4.A.1)*
Both $\mathbf{K}'\mathbf{M}\mathbf{K} = \mathbf{K}'(\mathbf{I}_{(n)} - \mathbf{R}\mathbf{R}')\mathbf{K} = \mathbf{I}_{(n-k)} - \mathbf{K}'\mathbf{R}\mathbf{R}'\mathbf{K}$ and $\mathbf{K}'\mathbf{R}\mathbf{R}'\mathbf{K}$ are nonnegative definite symmetric matrices, so that their eigenvalues $\lambda_i$ and $1-\lambda_i$, respectively, $i = 1, 2, \ldots, n-k$, are nonnegative. We assume the ordering $0 \leqslant \lambda_1 \leqslant \lambda_2 \leqslant \ldots \leqslant \lambda_{n-k} \leqslant 1$. In the same way, let $d_i$, $i = 1, 2, \ldots, k$, be the eigenvalues of $\mathbf{R}'\mathbf{P}\mathbf{P}'\mathbf{R} = \mathbf{R}'(\mathbf{I}_{(n)} - \mathbf{K}\mathbf{K}')$ $= \mathbf{I}_{(k)} - \mathbf{R}'\mathbf{K}\mathbf{K}'\mathbf{R}$, so that the $i$th eigenvalue of $\mathbf{R}'\mathbf{K}\mathbf{K}'\mathbf{R}$ is $1-d_i$, and $0 \leqslant d_1 \leqslant d_2 \leqslant \ldots \leqslant d_k \leqslant 1$. The nonzero eigenvalues of $\mathbf{K}'\mathbf{R}\mathbf{R}'\mathbf{K}$ are equal to those of $\mathbf{R}'\mathbf{K}\mathbf{K}'\mathbf{R}$ (see Result 1.A.6.4). Hence:

$$1-\lambda_i = 1-d_i \quad \text{for } i = 1, 2, \ldots, k$$
$$= 0 \quad \text{for } i = k+1, k+2, \ldots, n\text{-}k \quad \text{if } k < n\text{-}k$$

and:

$$1-\lambda_i = 1-d_i \quad \text{for } i = 1, 2, \ldots, n\text{-}k \qquad \text{if } k \geqslant n\text{-}k$$

In particular, the largest eigenvalue $1-\lambda_1$ of $\mathbf{K'RR'K}$ is equal to the largest eigenvalue $1-d_1$ of $\mathbf{R'KK'R}$, so that $\lambda_1 = d_1$. $(\mathbf{K'MK})^{-1/2}$ is defined if and only if all eigenvalues of $\mathbf{K'MK}$ are strictly positive, i.e. if and only if $\lambda_1 > 0$. Since $d_1$ is the smallest eigenvalue of $\mathbf{R'PP'R}$, we have $d_1 > 0$ if and only if the square matrix $\mathbf{P'R}$ is nonsingular. In view of $\lambda_1 = d_1$, (4.A.1) follows.

*Proof of (4.A.2)*
Using the relations between $\lambda_i$ and $d_i$, established in the proof of (4.A.1), we find when $k < n\text{-}k$ that:

$$tr(\mathbf{K'MK})^{1/2} = \sum_{i=1}^{n-k} \lambda_i^{1/2} = \sum_{i=1}^{k} d_i^{1/2} + n - 2k$$

and when $k \geqslant n\text{-}k$ that:

$$tr(\mathbf{K'MK})^{1/2} = \sum_{i=1}^{n-k} \lambda_i^{1/2} = \sum_{i=1}^{n-k} d_i^{1/2} = \sum_{i=1}^{k} d_i^{1/2} + n - 2k$$

where $n-2k = 0$ if $k = n\text{-}k$ and $1-d_{n\text{-}k+1} = 1-d_{n\text{-}k+2} = \ldots = 1-d_k = 0$ if $k > n\text{-}k$. Since the $d_i$ are the eigenvalues of $\mathbf{R'PP'R}$, and hence of $\mathbf{P'RR'P}$, (4.A.2) follows.

*Proof of (4.A.3)*
Let $\mathbf{L}$ be the matrix of eigenvectors of $\mathbf{P'RR'P}$, such that $\mathbf{P'RR'P} = \mathbf{LDL'}$, then:

$$\mathbf{R'PP'R} = (\mathbf{R'PLD}^{-1/2})\,\mathbf{D}\,(\mathbf{R'PLD}^{-1/2})'$$

where $(\mathbf{R'PLD}^{-1/2})'\,(\mathbf{R'PLD}^{-1/2}) = \mathbf{I}_{(k)}$, and we find:

$$\mathbf{R'KK'R} = \mathbf{I} - \mathbf{R'PP'R} = (\mathbf{R'PLD}^{-1/2})\,(\mathbf{I} - \mathbf{D})\,(\mathbf{R'PLD}^{-1/2})'$$

so that:

$$\mathbf{R}'\mathbf{K}\mathbf{K}'\mathbf{R}(\mathbf{R}'\mathbf{P}\mathbf{L}\mathbf{D}^{-1/2}) = (\mathbf{R}'\mathbf{P}\mathbf{L}\mathbf{D}^{-1/2})(\mathbf{I}-\mathbf{D})$$

Premultiplying this equation by $\mathbf{K}'\mathbf{R}$ and using $\mathbf{K}'\mathbf{R}\mathbf{R}'\mathbf{P} = -\mathbf{K}'\mathbf{M}\mathbf{P}$, since $\mathbf{K}'\mathbf{P} = \mathbf{0}$, we obtain:

$$\mathbf{K}'\mathbf{R}\mathbf{R}'\mathbf{K}\mathbf{F} = \mathbf{F}(\mathbf{I}-\mathbf{D})$$

where $\mathbf{F} = \mathbf{K}'\mathbf{M}\mathbf{P}\mathbf{L}\mathbf{D}^{-1/2}$. It is easily verified that $\mathbf{F}'\mathbf{F} = \mathbf{I}-\mathbf{D}$, which diagonal matrix is nonnegative definite – see the proof of (4.A.1). Suppose that $d_j = 1$, so that the $j$th column of $\mathbf{F}$ consists of zeros. Let us write $\boldsymbol{\Delta}$ for the diagonal matrix obtained from $\mathbf{D}$ by deleting row $j$ and column $j$ for every $j$ for which $d_j = 1$, and let $\mathbf{H}$ denote the matrix obtained from $\mathbf{F}$ by deleting column $j$ for every $j$ for which $d_j = 1$. Then we have $\mathbf{H}'\mathbf{H} = \mathbf{I}-\boldsymbol{\Delta}$ and:

$$[\mathbf{H}(\mathbf{I}-\boldsymbol{\Delta})^{-1/2}]'[\mathbf{H}(\mathbf{I}-\boldsymbol{\Delta})^{-1/2}] = \mathbf{I}$$

Returning to the above expression for $\mathbf{R}'\mathbf{K}\mathbf{K}'\mathbf{R}$, we obtain:

$$\begin{aligned}\mathbf{K}'\mathbf{R}(\mathbf{R}'\mathbf{K}\mathbf{K}'\mathbf{R})\mathbf{R}'\mathbf{K} &= \mathbf{K}'\mathbf{R}(\mathbf{R}'\mathbf{P}\mathbf{L}\mathbf{D}^{-1/2})(\mathbf{I}-\mathbf{D})(\mathbf{R}'\mathbf{P}\mathbf{L}\mathbf{D}^{-1/2})'\mathbf{R}\\ &= \mathbf{F}(\mathbf{I}-\mathbf{D})\mathbf{F}'\\ &= \mathbf{H}(\mathbf{I}-\boldsymbol{\Delta})\mathbf{H}'\\ &= [\mathbf{H}(\mathbf{I}-\boldsymbol{\Delta})^{-1/2}](\mathbf{I}-\boldsymbol{\Delta})^2[\mathbf{H}(\mathbf{I}-\boldsymbol{\Delta})^{-1/2}]'\end{aligned}$$

and hence:

$$\mathbf{K}'\mathbf{R}\mathbf{R}'\mathbf{K} = [\mathbf{H}(\mathbf{I}-\boldsymbol{\Delta})^{-1/2}](\mathbf{I}-\boldsymbol{\Delta})[\mathbf{H}(\mathbf{I}-\boldsymbol{\Delta})^{-1/2}]'$$

Let $\mathbf{G}$ denote a matrix such that:

$$[\mathbf{H}(\mathbf{I}-\boldsymbol{\Delta})^{-1/2} \vdots \mathbf{G}]' = [\mathbf{H}(\mathbf{I}-\boldsymbol{\Delta})^{-1/2} \vdots \mathbf{G}]^{-1}$$

Then:

$$\mathbf{K}'\mathbf{R}\mathbf{R}'\mathbf{K} = [\mathbf{H}(\mathbf{I}-\boldsymbol{\Delta})^{-1/2} \vdots \mathbf{G}]\begin{bmatrix}\mathbf{I}-\boldsymbol{\Delta} & \mathbf{0}\\ \mathbf{0} & \mathbf{0}\end{bmatrix}[\mathbf{H}(\mathbf{I}-\boldsymbol{\Delta})^{-1/2} \vdots \mathbf{G}]'$$

and:

$$\mathbf{K'MK} = [\mathbf{H}(\mathbf{I} - \mathbf{\Delta})^{-\frac{1}{2}} \vdots \mathbf{G}] \begin{bmatrix} \mathbf{\Delta} & \vdots & \mathbf{0} \\ \cdots & \vdots & \cdots \\ \mathbf{0} & \vdots & \mathbf{I} \end{bmatrix} [\mathbf{H}(\mathbf{I} - \mathbf{\Delta})^{-\frac{1}{2}} \vdots \mathbf{G}]'$$

Hence:

$$\begin{aligned} (\mathbf{K'MK})^{-\frac{1}{2}} &= [\mathbf{H}(\mathbf{I} - \mathbf{\Delta})^{-\frac{1}{2}} \vdots \mathbf{G}] \begin{bmatrix} \mathbf{\Delta}^{-\frac{1}{2}} & \vdots & \mathbf{0} \\ \cdots & \vdots & \cdots \\ \mathbf{0} & \vdots & \mathbf{I} \end{bmatrix} [\mathbf{H}(\mathbf{I} - \mathbf{\Delta})^{-\frac{1}{2}} \vdots \mathbf{G}]' \\ &= \mathbf{H}(\mathbf{I} - \mathbf{\Delta})^{-1} \mathbf{\Delta}^{-\frac{1}{2}} \mathbf{H}' + \mathbf{GG}' \\ &= \mathbf{H}(\mathbf{I} - \mathbf{\Delta})^{-1} \mathbf{\Delta}^{-\frac{1}{2}} \mathbf{H}' + \mathbf{I} - \mathbf{H}(\mathbf{I} - \mathbf{\Delta})^{-1} \mathbf{H}' \\ &= \mathbf{I} + \mathbf{H}(\mathbf{\Delta} + \mathbf{\Delta}^{\frac{1}{2}})^{-1} \mathbf{H}' \\ &= \mathbf{I} + \mathbf{F}(\mathbf{D} + \mathbf{D}^{\frac{1}{2}})^{-1} \mathbf{F}' \end{aligned}$$

which proves (4.A.3).

*Proof of (4.A.4)*
Substitution of (4.A.3) into $\mathbf{B}' = \mathbf{K}(\mathbf{K'MK})^{-\frac{1}{2}}\mathbf{K'M}$ and use of $\mathbf{L'P'MKK'M} = \mathbf{L'P'M} - \mathbf{L'P'MPP'M} = \mathbf{L'P'M} - (\mathbf{I} - \mathbf{D})\mathbf{L'P'M} = \mathbf{DL'P'M}$ yields:

$$\begin{aligned} \mathbf{B}' &= \mathbf{KK'M} + \mathbf{KK'MPLD}^{-1}(\mathbf{D} + \mathbf{D}^{\frac{1}{2}})^{-1}\mathbf{L'P'MKK'M} \\ &= \mathbf{M} - \mathbf{PP'M} + \mathbf{KK'V}(\mathbf{D} + \mathbf{D}^{\frac{1}{2}})^{-1}\mathbf{V}' \end{aligned}$$

where $\mathbf{V} = \mathbf{MPL}$. Then, using $\mathbf{P'M} = \mathbf{LV}'$ and $\mathbf{P'V} = \mathbf{P'MPL} = \mathbf{L}(\mathbf{I} - \mathbf{D})$, we obtain:

$$\begin{aligned} \mathbf{B}' &= \mathbf{M} - \mathbf{PLV}' + \mathbf{V}(\mathbf{D} + \mathbf{D}^{\frac{1}{2}})^{-1}\mathbf{V}' - \mathbf{PL}(\mathbf{I} - \mathbf{D})(\mathbf{D} + \mathbf{D}^{\frac{1}{2}})^{-1}\mathbf{V}' \\ &= \mathbf{M} + \mathbf{V}(\mathbf{D} + \mathbf{D}^{\frac{1}{2}})^{-1}\mathbf{V}' - \mathbf{PLD}^{-\frac{1}{2}}\mathbf{V}' \end{aligned}$$

Furthermore, with $\mathbf{U} = \mathbf{X}(\mathbf{X'X})^{-1}\mathbf{X'PL}$, we have:

$$\begin{aligned} \mathbf{I} - \mathbf{UD}^{-1}\mathbf{U}' &= \mathbf{I} - \mathbf{X}(\mathbf{X'X})^{-1}\mathbf{X'PLD}^{-1}\mathbf{L'P'X}(\mathbf{X'X})^{-1}\mathbf{X}' \\ &= \mathbf{I} - \mathbf{X}(\mathbf{X'X})^{-1}\mathbf{X'P}[\mathbf{P'X}(\mathbf{X'X})^{-1}\mathbf{X'P}]^{-1}\mathbf{P'X}(\mathbf{X'X})^{-1}\mathbf{X}' \\ &= \mathbf{M} \end{aligned}$$

so that:

$$\begin{aligned} \mathbf{B}' &= \mathbf{I} - \mathbf{U}\mathbf{D}^{-1}\mathbf{U}' + \mathbf{V}(\mathbf{D} + \mathbf{D}^{1/2})^{-1}\mathbf{V}' - (\mathbf{U} + \mathbf{V})\mathbf{D}^{-1/2}\mathbf{V}' \\ &= \mathbf{I} - \mathbf{U}\mathbf{D}^{-1}\mathbf{U}' - \mathbf{V}(\mathbf{I} + \mathbf{D}^{1/2})^{-1}\mathbf{V}' - \mathbf{U}\mathbf{D}^{-1/2}\mathbf{V}' \end{aligned}$$

The calculation of $\mathbf{V}$ and $\mathbf{U}$ may run along the sequence (given $\mathbf{X}$ anc $\mathbf{P}$): $\mathbf{X}'\mathbf{X}$, $(\mathbf{X}'\mathbf{X})^{-1}$, $\mathbf{X}'\mathbf{P}$, $(\mathbf{X}'\mathbf{X})^{-1}\mathbf{X}'\mathbf{P}$, $\mathbf{P}'\mathbf{X}(\mathbf{X}'\mathbf{X})^{-1}\mathbf{X}'\mathbf{P}$, $\mathbf{L}$ and $\mathbf{D}$, $(\mathbf{X}'\mathbf{X})^{-1}\mathbf{X}'\mathbf{PL}$, $\mathbf{X}(\mathbf{X}'\mathbf{X})^{-1}\mathbf{X}'\mathbf{PL} = \mathbf{U}$, $\mathbf{PL}$, $\mathbf{PL} - \mathbf{U} = \mathbf{V}$. Calculation of $\mathbf{B}'$ along this sequence and (4.A.4) reduced the required computer time by some 80 percent compared with a straightforward calculatic according to $\mathbf{B}' = \mathbf{K}(\mathbf{K}'\mathbf{MK})^{-1/2}\mathbf{K}'\mathbf{M}$ in applications to $15 \times 3$ $\mathbf{X}$-matrices.

*Proof of (4.A.5)*
From $\mathbf{B}' = \mathbf{M} + \mathbf{V}(\mathbf{D} + \mathbf{D}^{1/2})^{-1}\mathbf{V}' - \mathbf{PLD}^{-1/2}\mathbf{V}'$, see the proof of (4.A.4), we find:

$$\begin{aligned} \mathbf{w} &= \mathbf{B}'\mathbf{y} = \mathbf{My} + \mathbf{MPL}(\mathbf{D} + \mathbf{D}^{1/2})^{-1}\mathbf{L}'\mathbf{P}'\mathbf{My} - \mathbf{PLD}^{-1/2}\mathbf{L}'\mathbf{P}'\mathbf{M} \\ &= \hat{\mathbf{u}} + \mathbf{PL}[(\mathbf{D} + \mathbf{D}^{1/2})^{-1} - \mathbf{D}^{-1/2}]\mathbf{L}'\mathbf{P}'\hat{\mathbf{u}} - \mathbf{X}(\mathbf{X}'\mathbf{X})^{-1}\mathbf{X}'\mathbf{PL} \\ &\quad (\mathbf{D} + \mathbf{D}^{1/2})^{-1}\mathbf{L}'\mathbf{P}'\hat{\mathbf{u}} \end{aligned}$$

which expression is equal to (4.A.5). The calculation of $\mathbf{w}$ should no run along (4.A.4), because the calculation along (4.A.5) is much fast The calculation sequence may be (given $\mathbf{X}$, $\mathbf{P}$, and $\mathbf{y}$): $\mathbf{X}'\mathbf{X}$, $(\mathbf{X}'\mathbf{X})^{-1}$, $\mathbf{Xy}$, $(\mathbf{X}'\mathbf{X})^{-1}\mathbf{X}'\mathbf{y}$, $\mathbf{y} - \mathbf{X}(\mathbf{X}'\mathbf{X})^{-1}\mathbf{X}'\mathbf{y} = \hat{\mathbf{u}}$, $\mathbf{X}'\mathbf{P}$, $(\mathbf{X}'\mathbf{X})^{-1}\mathbf{X}'\mathbf{P}$, $\mathbf{P}'\mathbf{X}(\mathbf{X}'\mathbf{X})^{-1}\mathbf{X}'\mathbf{P}$, $\mathbf{L}$ and $\mathbf{D}$, $\mathbf{P}'\hat{\mathbf{u}}$, $\mathbf{L}'\mathbf{P}'\hat{\mathbf{u}}$, $(\mathbf{I} + \mathbf{D}^{1/2})^{-1}\mathbf{L}'\mathbf{P}'\hat{\mathbf{u}}$, $\mathbf{L}(\mathbf{I} + \mathbf{D}^{1/2})^{-1}\mathbf{L}'\mathbf{P}'\hat{\mathbf{u}}$, $\mathbf{PL}(\mathbf{I} + \mathbf{D}^{1/2})^{-1}\mathbf{L}'\mathbf{P}'\hat{\mathbf{u}} = \mathbf{a}$, $\mathbf{D}^{-1/2}(\mathbf{I} + \mathbf{D}^{1/2})^{-1}\mathbf{L}'\mathbf{P}'\hat{\mathbf{u}}$, $\mathbf{LD}^{-1/2}(\mathbf{I} + \mathbf{D}^{1/2})^{-1}\mathbf{L}'\mathbf{P}'\hat{\mathbf{u}}$, $(\mathbf{X}'\mathbf{X})^{-1}\mathbf{X}'\mathbf{PLD}^{-1/2}(\mathbf{I} + \mathbf{D}^{1/2})^{-1}\mathbf{L}'\mathbf{P}'\hat{\mathbf{u}}$, $\mathbf{X}(\mathbf{X}'\mathbf{X})^{-1}\mathbf{X}'\mathbf{PLD}^{-1/2}(\mathbf{I} + \mathbf{D}^{1/2})^{-1}\mathbf{L}'\mathbf{P}'\hat{\mathbf{u}} = \mathbf{b}$, $\hat{\mathbf{u}} - \mathbf{a} - \mathbf{b} = \mathbf{w}$.

# 5. Evaluation of the tests

## 5.1 Description of the test cases

In this chapter we compare powers of the tests against positive autocorrelation and against heterovariance, as described in Section 2.5, in order to answer the question of which disturbance estimator should be applied to which test. The six disturbance estimators, identified by $\hat{u}$, *hyp*, *sel*, *mod. BLUS*, *BLUS*, and *z*, are described below. The powers are calculated for $\mathcal{H}_A : \rho = 0.8$ in *test(Q)*, for $\mathcal{H}_A : \eta = 0.9$ in *test(S)*, and for $\mathcal{H}_A : \gamma = 0.83$ in *test(V)*, at significance levels $\alpha = 0.05$ and $\alpha = 0.10$ (we also carried out computations for $\mathcal{H}_A : \rho = 0.3$, or $\eta = 0.75$, or $\gamma = 0.5$, and $\alpha = 0.025$, the results of which are not presented in this study: we found the same relative performances of the disturbance estimators, so that the results of the power comparisons are not restricted to the rather arbitrary choice of $\mathcal{H}_A$ and $\alpha$). The **X**-matrices used in this chapter are also described below. The section is concluded by a comparison between the test statistics using *mod. BLUS* and those using *BLUS*.

Five of the six disturbance estimators (i.e. in all cases but *z*) are special cases of **w**:

$$\mathbf{w} = \mathbf{K}(\mathbf{K}'\mathbf{J}'\mathbf{M}\mathbf{J}\mathbf{K})^{-\frac{1}{2}}\mathbf{K}'\mathbf{J}'\mathbf{M}\mathbf{y} \tag{5.1}$$

where $\mathbf{KK}' = \mathbf{I} - \mathbf{PP}' = \mathbf{\Omega}$ is idempotent. In $\hat{u}$, *hyp*, *sel*, and *mod. BLUS*, **J** is equal to $\mathbf{I}_{(n)}$. In *BLUS*, **J** is equal to an $n \times (n\text{-}k)$ submatrix of $\mathbf{I}_{(n)}$: in *test(Q)* the last $k$ columns are deleted from $\mathbf{I}_{(n)}$, in the heterovariance tests the middle $k$ columns are deleted.

The estimator *z* is the **z** proposed by Durbin, see (3.4). The estimators *mod.BLUS*, *BLUS*, and *z* are applied only to $15 \times 3$ **X**-matrices. All six estimators are now determined as follows by **K**, or by **P**, which amounts to the same thing:

| | |
|---|---|
| $\hat{u}$ | $\mathbf{P} = \mathbf{R}$, so that $\Omega = \mathbf{M}$ and $\mathbf{w}$ is the *o.l.s.* residual vector $\hat{\mathbf{u}}$. |
| *hyp* | $\mathbf{P} = [\mathbf{h}_1^* : \mathbf{h}_2^* : \ldots : \mathbf{h}_k^*]$, see (4.8). |
| *sel* | $\mathbf{P} = [\mathbf{h}_{i1}^* : \mathbf{h}_{i2}^* : \ldots : \mathbf{h}_{ik}^*]$, see (4.9). |
| *mod. BLUS* | $\mathbf{P} = [\mathbf{e}_7 : \mathbf{e}_8 : \mathbf{e}_9]$, where $\mathbf{e}_i$ denotes the $i$th column of $\mathbf{I}_{(15)}$, see Section 3.3. |
| *BLUS* | $\mathbf{K} = \mathbf{I}_{(12)}$ (so that $\mathbf{P}$ does not exist). |
| *z* | $\mathbf{P}$ is identical to the $\mathbf{P}$ in *hyp*. |

The **P**-matrix in *mod. BLUS* is chosen in view of the heterovariance tests, and the same holds for **J** in *BLUS*. The **J**-matrix in *BLUS* when used in *test(Q)* is chosen for computational convenience.

In Section 5.2 experiments are carried out with **w** in (5.1), with $\mathbf{J} = \mathbf{I}_{(n)}$ and $\mathbf{P} = [\mathbf{h}_1^* : \mathbf{h}_{i2}^* : \mathbf{h}_{i3}^*]$, where $i2 = 2, 3, \ldots, 14$ and $i3 = i2+1, i2+2, \ldots, 15$. In Section 5.4 other experiments are carried out with **P**-matrices, whose orders differ from the order of **X**, where **J** plays the role of a *link matrix* between **M** and **K** in (5.1).

All the **X**-matrices contain a constant term column and one or more columns consisting of time series data taken from literature. The name of the **X**-matrix is followed by its order and a description of the time series data:

| | | |
|---|---|---|
| $\mathbf{X}_C$ | $15 \times 3$ | Chow (1957, logarithms of table 1); *log* automobile stock per capita and *log* personal money stock per capita for the United States, 1921-1935. |
| $\mathbf{X}_{C1}$ | $20 \times 3$ | *idem*; 1921-1940. |
| $\mathbf{X}_{C2}$ | $20 \times 4$ | *idem*; plus *log* expected income per capita, 1921-1940. |
| $\mathbf{X}_{C3}$ | $10 \times 3$ | *idem*; as $\mathbf{X}_{C1}$, 1921-1930. |
| $\mathbf{X}_{C4}$ | $10 \times 4$ | *idem*; as $\mathbf{X}_{C2}$, 1921-1930. |
| $\mathbf{X}_H$ | $16 \times 5$ | Henshaw (1966, table 1); supply of California, Oregon, and Washington pears and United States non-agricultural income, 1925-1940. |
| $\mathbf{X}_K$ | $15 \times 3$ | Klein (1950, p. 135); profits and wages for the United States, 1923-1937. |
| $\mathbf{X}_{K1}$ | $20 \times 3$ | *idem*; 1921-1940. |
| $\mathbf{X}_{K2}$ | $10 \times 3$ | *idem*; 1921-1930. |
| $\mathbf{X}_{K3}$ | $10 \times 2$ | *idem*; wages only, 1921-1930. |

| | | |
|---|---|---|
| $\mathbf{X}_S$ | $15 \times 3$ | Sato (1970, p. 203); capital and man-hours for the United States, 1946-1960. |
| $\mathbf{X}_T$ | $15 \times 3$ | Theil (1971, table 3.1); *log* real per capita income and *log* relative price of textiles for the Netherlands, 1923-1937. |
| $\mathbf{X}_{T1}$ | $17 \times 3$ | *idem*; 1923-1939. |
| $\mathbf{X}_{T2}$ | $12 \times 3$ | *idem*; 1923-1934. |
| $\mathbf{X}_A$ | $15 \times 3$ | Koerts and Abrahamse (1969, pp. 153-154); two columns of artificial data. |

The regressor data are chosen partly because of their use as test cases by other authors (Durbin 1970: Klein and Theil data; Durbin and Watson 1971: Klein, Theil, and Henshaw data; Koerts and Abrahamse 1969: Theil and artificial data), and partly arbitrarily (Chow and Sato data).

In Section 3.3 we mentioned the fact that the *BLUS* estimator, say $\mathbf{w}_1 = \mathbf{B}_1'\mathbf{y}$, is equal to the *modified BLUS* estimator, say $\mathbf{w}_2 = \mathbf{B}_2'\mathbf{y}$, apart from $k$ zeros, if $\mathbf{K}$ in *mod. BLUS* is equal to $\mathbf{J}$ in *BLUS*. Taking $\mathbf{P}$ in *mod. BLUS* and $\mathbf{J}$ in *BLUS* as specified above (where $n = 15$ and $k = 3$), we have the following situation in the heterovariance tests: the twelve columns of the $15 \times 12$ matrix $\mathbf{B}_1$ are equal to the first six and the last six columns of the $15 \times 15$ matrix $\mathbf{B}_2$, while the middle three columns of $\mathbf{B}_2$ consist of zeros. It follows that $\mathbf{B}_1\mathbf{B}_1' = \mathbf{B}_2\mathbf{B}_2'$.

In *test(V)*, see (2.20), the distribution function of $V$ is determined by the eigenvalues of $\mathbf{SB}[(1+\nu)\mathbf{A}_\nu - \nu\mathbf{I}_{(p)}]\mathbf{B}'\mathbf{S}'$, see (2.13), where $p = 12$ if $\mathbf{B} = \mathbf{B}_1$, and $p = 15$ if $\mathbf{B} = \mathbf{B}_2$, and where $\mathbf{SS}' = \mathbf{\Gamma}$. Using $\mathbf{B}_1\mathbf{B}_1' = \mathbf{B}_2\mathbf{B}_2'$ and $\mathbf{B}_1\mathbf{A}_{\nu(12)}\mathbf{B}_1' = \mathbf{B}_2\mathbf{A}_{\nu(15)}\mathbf{B}_2'$, where:

$$\mathbf{A}_{\nu(12)} = \begin{bmatrix} \mathbf{I}_{(6)} & \vdots & 0 \\ \cdots & \vdots & \cdots \\ 0 & \vdots & 0_{(6)} \end{bmatrix} \quad \text{and } \mathbf{A}_{\nu(15)} = \begin{bmatrix} \mathbf{I}_{(7)} & \vdots & 0 \\ \cdots & \vdots & \cdots \\ 0 & \vdots & 0_{(8)} \end{bmatrix}$$

we find $\mathbf{SB}_1[(1+\nu)\mathbf{A}_\nu - \nu\mathbf{I}_{(12)}]\mathbf{B}_1'\mathbf{S}' = \mathbf{SB}_2[(1+\nu)\mathbf{A}_\nu - \nu\mathbf{I}_{(15)}]\mathbf{B}_2'\mathbf{S}'$ for all $\nu$ and all $\mathbf{S}$. Hence, the null distribution and the alternative distribution of $V$ using *BLUS* are identical to the null distribution and the alternative distribution of $V$ using *mod. BLUS*, respectively. For other $\mathbf{K}$-matrices in *mod. BLUS* one may find that only the null distributions or that none of the distributions are identical.

In *test(S)*, see (2.19), the distribution function of $S$ is determined

by the eigenvalues of $\mathbf{SB}(\mathbf{A}_s - s\mathbf{I}_{(p)})\mathbf{B}'\mathbf{S}'$. Here we cannot establish an equality or a proportionality between the eigenvalues when $\mathbf{B} = \mathbf{B}_1$ on the one hand and when $\mathbf{B} = \mathbf{B}_2$ on the other hand. The calculations of significance points and powers prove that the distributions are not identical. If $\mathbf{K}$ in *mod. BLUS* were defined as consisting of the first twelve columns from $\mathbf{I}_{(15)}$, then we would have a fixed proportionality between the significance points $s_1$ and $s_2$ of $S$ using *BLUS* and *mod. BLUS*, respectively. This can be seen as follows. The eigenvalues of $\mathbf{K}'(\mathbf{A}_s - s\mathbf{I}_{(p)})\mathbf{K}$ determine $\mathcal{F}(s\,|\mathcal{H}_0)$, see Section 2.6. Taking $\mathbf{K}$ as suggested above, the null distribution of $S$ using *mod. BLUS* is determined by the eigenvalues $i/15 - s_2$ for $i = 1, 2, \ldots, 12$. For all *BLUS* vectors we have $\mathbf{K} = \mathbf{I}_{(12)}$, so that the null distribution of $S$ using *BLUS* is determined by the eigenvalues $i/12 - s_1$ for $i = 1, 2, \ldots, 12$. Hence, $\mathcal{F}(s_1\,|\mathcal{H}_0) = Pr[\sum_{i=1}^{12} (i/12 - s_1)z_i^2 \leqslant 0]$ and $\mathcal{F}(s_2\,|\mathcal{H}_0) = Pr[\sum_{i=1}^{12} (i/15 - s_2)z_i^2 \leqslant 0]$, so that $\alpha = \mathcal{F}(s_1\,|\mathcal{H}_0) = \mathcal{F}(s_2\,|\mathcal{H}_0)$ for all values of $\alpha$ implies $s_1/s_2 = 15/12$.
In *test(Q)*, both the null distribution and the alternative distribution of $Q$ using *BLUS* always differ from those distributions of $Q$ using *mod. BLUS*.

## 5.2 Values of $\phi$ and the selection device

Given $\mathbf{X}$, the values of $\phi$, see (4.3), can be calculated for all $\mathbf{P}$-matrices under consideration. In the case of $\hat{u}$, where $\mathbf{P} = \mathbf{R}$, we have $\phi = 1$ and in the case of *BLUS* the measure is not defined. Since $\mathbf{P}$ in *hyp* and $\mathbf{P}$ in $z$ are identical, we have the same values of $\phi$ in these two cases. In the case of *sel* the indices $i1, i2, \ldots, ik$ in $\mathbf{P} = [\mathbf{h}_{i1}^* : \mathbf{h}_{i2}^* : \ldots : \mathbf{h}_{ik}^*]$ depend on $\mathbf{X}$ in accordance with (4.9). In Table 5.1 we see that the values of $\phi$ in the case of *mod. BLUS* are very small. The average value of $\phi$ in the case of *hyp* (and $z$) is 0.75, which is in agreement with our findings in Section 4.6. The values of $\phi$ in the case of *sel* are at least as great as in the case of *hyp*, with average value 0.83. All $\mathbf{P}$-matrices in *sel*, corresponding to empirical $\mathbf{X}$-matrices, include both $\mathbf{h}_1^*$ and $\mathbf{h}_2^*$. Usually, $\mathbf{P}$ includes $\mathbf{h}_3^*$ if $n \leqslant 16$ and $\mathbf{h}_4^*$ if $n > 16$. The vectors $\mathbf{h}_1^*$ and $\mathbf{h}_2^*$ represent a constant term and an almost linear trend, respectively, irrespective of $n$. But the period of the waves, represented by $\mathbf{h}_3^*$, $\mathbf{h}_4^*$, and so on, depends on $n$. For instance, $\mathbf{h}_3^*$ if $n = 12$ and $\mathbf{h}_4^*$ if $n = 18$ represent waves with the same period. Hence, when a given $\mathbf{X}$-matrix

*Table 5.1* The measure $\phi$ and the selection device.

| Matrix | order | Values of $\phi$ | | | Selected indices | | | | |
|---|---|---|---|---|---|---|---|---|---|
| $\mathbf{X}$ | $n \times k$ | *mod.BLUS* | *hyp/z* | *sel* | *i*1 | *i*2 | *i*3 | *i*4 | *i*5 |
| $\mathbf{X}_{K3}$ | 10 x 2 | | 0.921 | 0.921 | 1 | 2 | | | |
| $\mathbf{X}_{C3}$ | 10 x 3 | | 0.666 | 0.750 | 1 | 2 | 8 | | |
| $\mathbf{X}_{K2}$ | 10 x 3 | | 0.803 | 0.803 | 1 | 2 | 3 | | |
| $\mathbf{X}_{C4}$ | 10 x 4 | | 0.741 | 0.817 | 1 | 2 | 3 | 8 | |
| $\mathbf{X}_{T2}$ | 12 x 3 | | 0.938 | 0.938 | 1 | 2 | 3 | | |
| $\mathbf{X}_{C}$ | 15 x 3 | 0.169 | 0.853 | 0.853 | 1 | 2 | 3 | | |
| $\mathbf{X}_{K}$ | 15 x 3 | 0.172 | 0.605 | 0.844 | 1 | 2 | 4 | | |
| $\mathbf{X}_{S}$ | 15 x 3 | 0.090 | 0.677 | 0.779 | 1 | 2 | 8 | | |
| $\mathbf{X}_{T}$ | 15 x 3 | 0.217 | 0.887 | 0.887 | 1 | 2 | 3 | | |
| $\mathbf{X}_{A}$ | 15 x 3 | 0.232 | 0.767 | 0.873 | 1 | 3 | 4 | | |
| $\mathbf{X}_{H}$ | 16 x 5 | | 0.562 | 0.652 | 1 | 2 | 3 | 6 | 16 |
| $\mathbf{X}_{T1}$ | 17 x 3 | | 0.760 | 0.819 | 1 | 2 | 4 | | |
| $\mathbf{X}_{C1}$ | 20 x 3 | | 0.686 | 0.843 | 1 | 2 | 4 | | |
| $\mathbf{X}_{K1}$ | 20 x 3 | | 0.648 | 0.845 | 1 | 2 | 4 | | |
| $\mathbf{X}_{C2}$ | 20 x 4 | | 0.707 | 0.853 | 1 | 2 | 4 | 5 | |

with $n = 15$ is enlarged such that $n = 20$, and the selection device selects $\mathbf{h}_3^*$ in the $n = 15$ case, then it is not amazing that the selection device selects $\mathbf{h}_4^*$ in the $n = 20$ case, especially when the enlargement takes place at one end of the matrix. This occurs in the cases where $\mathbf{X}_C$ is enlarged to $\mathbf{X}_{C1}$ and $\mathbf{X}_{C2}$, and where $\mathbf{X}_{T2}$ is enlarged to $\mathbf{X}_{T1}$. The enlargement from $\mathbf{X}_K$ to $\mathbf{X}_{K1}$ takes place at both ends of the matrix, but here the selection for $\mathbf{X}_K$ already includes $\mathbf{h}_4^*$. In view of these selections, a shift from $\mathbf{h}_3^*$ to $\mathbf{h}_4^*$ seems to be regular. In Louter and Dubbelman (1973) other series are analyzed and there such a shift is completely absent, however.[1]

1. They used three $\mathbf{X}$-matrices, each containing a constant term and Dutch data from 1953 to 1972 on either total imports, or private and public consumption, or average nominal wage rate and employed labour and unemployed labour, thus yielding a 20 x 2, a 20 x 3 and a 20 x 4 matrix. Submatrices are obtained by taking the observation periods 1953 to 1967 and 1961 to 1970, so that $n = 15$ and 10, respectively. The selection device indicates $\mathbf{P} = [\mathbf{h}_1^* : \mathbf{h}_2^*]$ for the $n \times 2$ matrices with $n = 10$, 15, and 20, and $\mathbf{P} = [\mathbf{h}_1^* : \mathbf{h}_2^* :$ for the $n \times 3$ matrices with $n = 10$, 15, and 20, and $\mathbf{P} = [\mathbf{h}_1^* : \mathbf{h}_2^* : \mathbf{h}_3^* : \mathbf{h}_i^*]$ for the $n \times 4$ matrices with $n = 10$ (where $i = 4$), 15 (where $i = 5$), and 20 (where $i = 9$).

For each of the five 15 × 3 **X**-matrices in Table 5.1 we computed the value of the measure $\psi$, see (4.2), for all admissible **P**-matrices (i.e. 15 × 3 matrices consisting of $\mathbf{h}_1^*$ and two of $\mathbf{h}_2^*, \mathbf{h}_3^*, \ldots, \mathbf{h}_{15}^*$). The **P** for which $\psi$ is maximal with respect to **X** coincides with the **P** indicated by the selection device (maximal $\phi$), in all five cases.

This supports the switch from the criterion of maximal $\psi$, derived from optimization of the *best* estimation criterion, to the much more convenient criterion of maximal $\phi$.

For $\mathbf{X}_C$ and $\mathbf{X}_T$ we computed the power of *test(Q)* at significance level 0.05 for all admissible **P**-matrices. In the case of $\mathbf{X}_T$ the highest power is scored for $\mathbf{P} = [\mathbf{h}_1^* : \mathbf{h}_2^* : \mathbf{h}_3^*]$. For $\mathbf{X} = \mathbf{X}_C$ the highest power scores are presented in Table 5.2, together with the measure $\phi$ and powers of *test(S)* at the same significance level.

*Table 5.2* The measure $\phi$ and powers of *test(Q)* and *test(S)* at $\alpha = 0.05$ for $\mathbf{X} = \mathbf{X}_C$ and $\mathbf{P} = [\mathbf{h}_1^* : \mathbf{h}_{i2}^* : \mathbf{h}_{i3}^*]$, where $\mathcal{H}_A : \rho = 0.8$ and $\eta = 0.9$, respectively.

| *i*2 | *i*3 | $\phi$ | Power of *test(Q)* | Power of *test(S)* |
|---|---|---|---|---|
| 4 | 8 | 0.406 | 0.566 | 0.290 |
| 3 | 8 | 0.668 | 0.563 | 0.339 |
| 5 | 8 | 0.382 | 0.563 | 0.226 |
| 6 | 8 | 0.395 | 0.556 | 0.247 |
| | $\hat{u}$ | 1.000 | 0.536 | 0.407 |
| 3 | 4 | 0.663 | 0.534 | 0.339 |
| 7 | 8 | 0.374 | 0.525 | 0.257 |
| 4 | 6 | 0.390 | 0.522 | 0.252 |
| 4 | 5 | 0.377 | 0.521 | 0.302 |
| 2 | 8 | 0.596 | 0.518 | 0.355 |

Table 5.2 reveals several facts. The measure $\phi$ is not a general power indicator in the sense that maximal power corresponds to maximal $\phi$, at least when $\mathbf{X} = \mathbf{X}_C$. In particular, the maximum value of $\phi$ is 1, which corresponds to the tests using $\hat{u}$. The exact Durbin-Watson test, i.e. *test(Q)* using $\hat{u}$, is overpowered four times, by adopting **P**-matrices with small $\phi$-values in connection with $\mathbf{X}_C$. However this does not imply that the selection device becomes useless. The purpose of Table 5.2 is to illustrate that our maximal $\phi$-approach has

some shortcomings. One of these shortcomings is that **P** is chosen for a given **X**, regardless of the test. Table 5.2 shows that the best **P** for $\mathbf{X} = \mathbf{X}_C$ in *test(Q)* is not the best **P** for $\mathbf{X} = \mathbf{X}_C$ in *test(S)*. These shortcomings should not be exaggerated, as we shall see in the next section.

## 5.3 Evaluation of the disturbance estimators in *test (Q)*

In Table 5.3 we present the powers of *test(Q)* using one of the six disturbance estimators, applied to the five 15 × 3 **X**-matrices, see Section 5.1. *Test (Q)* using $\hat{u}$ (i.e. the exact Durbin-Watson test) is not tabulable. Its powers have been included in Table 5.3 for the purpose of power comparison, because this test is close to *UMP*, see (2.5).

*Table 5.3* Powers (× 1000) at significance level $\alpha$ of the autocorrelation *test(Q)* using the six disturbance estimators, applied to the five 15 × 3 **X**-matrices ($\mathcal{H}_A : \rho = 0.8$).

| | $\mathbf{X}_C$ | $\mathbf{X}_K$ | $\mathbf{X}_S$ | $\mathbf{X}_T$ | $\mathbf{X}_A$ | Average |
|---|---|---|---|---|---|---|
| | | | $\alpha = 0.05$ | | | |
| $\hat{u}$ | 536 | 629 | 628 | 508 | 628 | 586 |
| *hyp* | 466 | 605 | 355 | 499 | 626 | 510 |
| *sel* | 466 | 610 | 625 | 499 | 628 | 566 |
| *z* | 469 | 551 | 350 | 344 | 577 | 458 |
| *mod. BLUS* | 424 | 501 | 436 | 360 | 532 | 451 |
| *BLUS* | 422 | 423 | 569 | 414 | 485 | 463 |
| | | | $\alpha = 0.10$ | | | |
| $\hat{u}$ | 665 | 734 | 738 | 633 | 726 | 699 |
| *hyp* | 595 | 715 | 479 | 625 | 725 | 628 |
| *sel* | 595 | 719 | 734 | 625 | 726 | 680 |
| *z* | 595 | 679 | 478 | 485 | 694 | 586 |
| *mod. BLUS* | 564 | 631 | 586 | 510 | 646 | 587 |
| *BLUS* | 562 | 568 | 693 | 556 | 613 | 598 |

*Test (Q)* using *sel* emerges as the best tabulable test, almost as powerful as the exact Durbin-Watson test in four of the five cases. The second best disturbance estimator is *hyp*, with a very low power score for *test(Q)* using *hyp* in $\mathbf{X} = \mathbf{X}_S$. The three remaining estimators are, on average, equally bad compared with *sel*.

We continue to consider $\hat{u}$, *hyp*, and *sel* in tests applied to **X**-matrices with different orders. The powers are presented in Table 5.4. Again *test(Q)* using *sel* is almost as good as *test (Q)* using $\hat{u}$, apart from one case, namely $\mathbf{X} = \mathbf{X}_{K2}$, while *test(Q)* using *hyp*

*Table 5.4* Powers (× 1000) at significance level $\alpha$ of *test(Q)* using $\hat{u}$, *hyp*, or *sel*, applied to other $n \times k$ **X**-matrices ($\mathcal{H}_A$ : $\rho = 0.8$).

| **X** | $\mathbf{X}_{C4}$ | $\mathbf{X}_{K2}$ | $\mathbf{X}_{C3}$ | $\mathbf{X}_{K3}$ | $\mathbf{X}_{T2}$ | $\mathbf{X}_{T1}$ | $\mathbf{X}_{C2}$ | $\mathbf{X}_{C1}$ | $\mathbf{X}_{K1}$ |
|---|---|---|---|---|---|---|---|---|---|
| $n \times k$ | 10x4 | 10x3 | 10x3 | 10x2 | 12x3 | 17x3 | 20x4 | 20x3 | 20x3 |
| | | | | | $\alpha = 0.05$ | | | | |
| $\hat{u}$ | 216 | 285 | 400 | 401 | 355 | 628 | 714 | 764 | 786 |
| *hyp* | 107 | 216 | 152 | 393 | 351 | 615 | 691 | 749 | 762 |
| *sel* | 207 | 216 | 390 | 393 | 351 | 624 | 697 | 749 | 769 |
| | | | | | $\alpha = 0.10$ | | | | |
| $\hat{u}$ | 330 | 421 | 527 | 533 | 483 | 733 | 803 | 845 | 861 |
| *hyp* | 183 | 343 | 238 | 525 | 479 | 723 | 785 | 834 | 845 |
| *sel* | 320 | 343 | 521 | 525 | 479 | 730 | 789 | 833 | 849 |

seems to be more unreliable for smaller $n$.

With regard to the required computing time, we compared *test(Q)* using *sel* and the *d*-test consisting of the bounds test and the beta approximation, see Section 2.7. Note that the calculation of a test statistic requires an observation vector **y**. We found that, for $n = 15$ and $k = 3$, a complete *d*-test procedure is slightly faster (less than 10 percent) than the calculation of $Q$ using *sel*, including the selection procedure. For our five $15 \times 3$ **X**-matrices the exact significance levels, corresponding to the beta-approximated 5 percent significance points, range from 0.0486 to 0.0507. The significance points are presented in Table 5.5.

Our conclusion is that *test(Q)* using *sel* can be recommended when an exact test against positive autocorrelation is required. If the test does not have to be exact, the choice between this test and the *d*-test with the beta approximation is optional.

*Table 5.5* Exact ($q$) and beta-approximated ($q_a$) 5 percent significance points.

| | $x_C$ | $x_K$ | $x_S$ | $x_T$ | $x_A$ |
|---|---|---|---|---|---|
| $q$ | 1.487 | 1.463 | 1.356 | 1.512 | 1.495 |
| $q_a$ | 1.481 | 1.467 | 1.353 | 1.511 | 1.496 |

In Louter and Dubbelman (1973) a complete Fortran computer program for *test(Q)* using *sel* is submitted. The program includes the selection device and the generation of **P**. A table of 5 and 10 percent significance points is given for $9 \leqslant n \leqslant 20$ and $k \leqslant 4$. The program also includes the computation of the critical level $\hat{\alpha}$ of the test. The critical level is the smallest significance level at which $\mathcal{H}_0$ would be rejected for the given **X** and **y**, see Section 2.3. Judging a testing result by means of a comparison between the critical level $\hat{\alpha}$ and the significance level $\alpha$ is preferable to that by means of a comparison between the value of the test statistic $Q$ and the significance point $q$. The point is that $q$ must be found by consulting a table of significance points, while $\alpha$ is a value which one has in mind. Besides, $\hat{\alpha}$ is much more informative, since its value can be compared with all significance levels $\alpha$, whereas tables of significance points are available for a small number of significance levels. The additional computer time for the calculation of $\hat{\alpha}$ is not very great if **P** consist of **h***-vectors, which is the case in *sel* (and also in *hyp* and $z$): on an IBM1130 computer with hardware floating point the time necessary for the calculation of $Q$ using *sel* is ¾ of a second ($n = 15, k = 3$), and the calculation of $\hat{\alpha}$ requires one additional second. If **K** does not consist of **h***-vectors, then the eigenvalues, which determine the probability distribution of the test statistic, see (2.15), cannot be generated by (2.22) – here we note that the calculation of a significance point for $Q$ using $\hat{u}$ takes some 40 seconds on the same computer if the $15 \times 3$ **X**-matrix contains a constant term, and some 120 seconds if that matrix does not contain a constant term. In the first case we have $\lambda_i(q) = \lambda_i(0) - q$, where the $\lambda_i(0)$ are the eigenvalues of $\mathbf{MA}_d\mathbf{M}$, apart from $k$ zeros; in the second case the $\lambda_i(q)$ are the eigenvalues of $\mathbf{M}(\mathbf{A}_d - q\mathbf{E})\mathbf{M}$, apart from $k$ zeros, see Section 2.6).

For the purpose of making a balanced choice between the significance level and the power, we performed least-squares fits of $\alpha$ and several functions of $n$ and $k$ on powers of *test(Q)* using $\hat{u}$. Data from 23 **X**-matrices (5 from Table 5.3, 9 from Table 5.4, and 9 from Louter and Dubbelman 1973) yielded the following relation between the power $p_{\mathbf{X}}$ and $n$, $k$, and $\alpha$ (with correlation coefficient $R^2$ = 0.975 and standard errors between brackets).

$$p_{\mathbf{X}} = \underset{(0.04)}{0.30} + \underset{(0.26)}{2.17\alpha} + \underset{(0.002)}{0.024(n+k)} - \underset{(0.11)}{1.35\frac{k}{n}} + \nu$$

where $\nu$ is a regression residual. On average, $\nu = -0.02$ for **X**-matrices where *sel* and *hyp* coincide, and $\nu = 0.02$ for **X**-matrices where *sel* and *hyp* differ.

Some of our observations tempt to hypothesize that the power $p_{\mathbf{X}}$ of the exact Durbin-Watson test for any $n \times k$ matrix **X** lies between two bounds, given $\alpha$. Indeed we found for $\alpha = 0.05$ and $\alpha = 0.10$ that $p_u \leqslant p_{\mathbf{X}} \leqslant p_l$ for all **X**-matrices with $n$ = 10, 15, 20 and $k$ = 3, dealt with in this chapter, where $p_u$ and $p_l$ are the powers for $\mathbf{X} = \mathbf{X}_u = [\mathbf{h}_1^* : \mathbf{h}_2^* : \mathbf{h}_3^*]$ and for $\mathbf{X} = \mathbf{X}_l = [\mathbf{h}_1^* : \mathbf{h}_{n-1}^* : \mathbf{h}_n^*]$, respectively. $\mathbf{X}_u$ and $\mathbf{X}_l$ are the matrices corresponding to the upper bound and the lower bound of the Durbin-Watson bounds test, see Section 2.7. Perhaps the hypothesis can be proved analytically under the conditions where the exact Durbin-Watson test is *UMPS*, see Section 2.1.

## 5.4 Experiments with the matrix J

Formula (5.1) allows for an $n \times p$ matrix **J** as a link matrix between the $n \times n$ matrix **M** with rank *n-k* and the $p \times r$ matrix **K**, with rank $r$, $r \leqslant n-k$. In our experiments we take **J** equal to the unit matrix supplemented with *p-n* zero columns if $p-n > 0$, or with *n-p* zero rows if $n-p > 0$.

In *hyp*, **K** consists of the last *n-k* columns of the $n \times n$ matrix **H***. We identify the estimator **w** with **J** given above and with **K** consisting of the last $r$ columns of the $p \times p$ matrix **H*** by *adapted hyp*. Significance points of *test(Q)* using *adapted hyp* applied to an $n \times k$ **X**-matrix can thus be found in the table of significance points of *test(Q)* using *hyp* (i.e. the table of the upper bound), not at the entries $n$ and $k$, but at the entries $p$ and $p$-$r$, since the distribution of $Q$ now depends on the $p \times (p\text{-}r)$matrix $\mathbf{P} = [\mathbf{h}_1^* \vdots \mathbf{h}_2^* \vdots \ldots \vdots \mathbf{h}_{p-r}^*]$. When $r = p$, the matrix **K** is square, like **K** in *BLUS*. Then the significance point is the point corresponding to the $p \times 1$ matrix $\mathbf{P} = \mathbf{h}_1^*$, see $\lambda_i(q)$ in Section 2.6.

In Table 5.6 powers of *test(Q)* using *adapted hyp* are presented.

*Table 5.6* Powers (× 1000) at significance level 0.05 of *test(Q)* using *adapted hyp* ($\mathcal{H}_A$ : $\rho = 0.8$).

| **X** n × k | r \ p | 11 | 12 | 13 | 14 | 15 | 16 | 17 | 18 |
|---|---|---|---|---|---|---|---|---|---|
| $\mathbf{X}_C$ 15 × 3 | 10 | 401 | 414 | 362 | 288 | 230 | 199 | 144 | 148 |
| | 11 | 404 | 437 | 412 | 415 | 307 | 262 | 170 | 136 |
| | 12 | | *422* | 417 | 446 | *466* | 328 | 201 | 151 |
| $\mathbf{X}_T$ 15 × 3 | 10 | 447 | 383 | 389 | 310 | 269 | 242 | 117 | 117 |
| | 11 | 435 | 436 | 377 | 428 | 368 | 298 | 192 | 149 |
| | 12 | | *414* | 422 | 469 | *499* | 378 | 256 | 182 |
| $\mathbf{X}_{T1}$ 17 × 3 | 10 | 542 | 384 | 392 | 311 | 259 | 195 | 168 | 175 |
| | 11 | 522 | 540 | 362 | 430 | 361 | 282 | 238 | 215 |
| | 12 | | 506 | 542 | 460 | 494 | 392 | 319 | 312 |
| | 13 | | | 457 | 524 | 524 | 548 | 432 | 409 |
| | 14 | | | | *435* | 512 | 576 | *615* | 489 |

There we italicized the powers of the test using *BLUS*(0.422, 0.414, 0.435) and of the test using *hyp*(0.466, 0.499, 0.615). For all three **X**-matrices, the power for *hyp* is greater than the powers for all other *adapted hyp* cases. When moving from the power for *hyp* upwards and to the right, the power decreases rapidly, while we observe a much slower decrease when moving to the left. In particular, fixing $r = n$-$k$, the loss of power is not dramatic when moving from $p = n$ to $p = n$-1, i.e. moving from the $15 \times 3$ matrix $\mathbf{P} = [\mathbf{h}_1^* \vdots \mathbf{h}_2^* \vdots \mathbf{h}_3^*]$ to the $14 \times 2$ matrix $\mathbf{P} = [\mathbf{h}_1^* \vdots \mathbf{h}_2^*]$ in the case of the $15 \times 3$ **X**-matrices. Note that the powers for $\boldsymbol{\Omega} = \mathbf{E}_{(p)}$ (where $p \leqslant n$-$k$ and $r = p$-1) are greater than the powers for $\boldsymbol{\Omega} = \mathbf{I}_{(p)}$ (where $r = p$) in all cases but one, namely for $\mathbf{X} = \mathbf{X}_C$ and $p = 11$. The significance point for $\boldsymbol{\Omega} = \mathbf{E}_{(p)}$ is equal to the significance point for $\boldsymbol{\Omega} = \mathbf{I}_{(p)}$. The powers for

*BLUS*, where $\Omega = \mathbf{I}_{(n-k)}$, are improved by taking $\Omega = \mathbf{E}_{(n-k)}$ in all three cases. Nevertheless, the best choice of $p$ and $r$ seems to be $p = n$ and $r = n\text{-}k$, i.e. *hyp*.

By *adapted sel* we identify the estimator **w** in (5.1) with **J** given in the first paragraph of this section and with **K** specified according to an adapted selection device. Whereas the selection device (4.9) prescribes **K** to consist of the **h***-vectors corresponding to the greatest $n$-$k$ diagonal elements of $\mathbf{H}^{*\prime}\mathbf{M}\mathbf{H}^*$, it is natural that a more general selection device regards the diagonal of $\mathbf{H}^{*\prime}\mathbf{J}'\mathbf{M}\mathbf{J}\mathbf{H}^*$: the greatest $r$ diagonal elements indicate the **h***-vectors to be included in the $p \times r$ matrix **K**. We applied *test(Q)* using *adapted sel* only once, namely in the case of the $16 \times 5$ matrix $\mathbf{X}_H$. Compared with $\hat{u}$, we find a great loss of power when *hyp* and *adapted hyp* are used, and acceptable powers when *sel* and *adapted sel* are used, see Table 5.7.

*Table 5.7* **Powers at significance level 0.05 of *test(Q)* applied to the 16 x 5 matrix $\mathbf{X}_H$.**

| Disturbance estimator | Selected indices *i*1 | *i*2 | *i*3 | *i*4 | *i*5 | Power |
|---|---|---|---|---|---|---|
| $\hat{u}$ | | | | | | 0.503 |
| *hyp* | 1 | 2 | 3 | 4 | 5 | 0.126 |
| *sel* | 1 | 2 | 3 | 6 | 16 | 0.441 |
| *adapted hyp* $p = 15, r = 11$ | 1 | 2 | 3 | 4 | | 0.174 |
| *adapted sel* $p = 15, r = 11$ | 1 | 2 | 3 | 14 | | 0.451 |

Of course, these experiments do not permit definite conclusions to be drawn. The purpose of this section is merely to indicate possible applications of a $p$-element vector **w** with $p \neq n$ and $p \neq n\text{-}k$. Perhaps *adapted sel* is attractive for **X**-matrices with $k > 4$ and with $p$ and $r$ such that $p\text{-}4 = r = n\text{-}k$, so that **P** consists of four columns. In all events, the use of **J** may reduce the required table of significance points and it even may entail a gain of power compared with *sel*, as in the case of $\mathbf{X}_H$.

## 5.5 Evaluation of the disturbance estimators in *test(S)* and *test(V)*

An impression of the merits of the six disturbance estimators, when used in the heterovariance *tests(S)* and *(V)*, is given in Tables 5.8 and 5.9. *Test(V)* using *mod. BLUS* and *test(V)* using *BLUS* are identical, see Section 5.1. Compared with the power scores of the tests using $\hat{u}$, the powers of the tests using *hyp* are very satisfactory. When used in *test(S)*, *sel* results in a loss of power compared with *hyp* at $\alpha = 0.05$ if $\mathbf{X} = \mathbf{X}_A$ only, *mod. BLUS* and *BLUS* are less powerful in all cases, and *z* appears to be unreliable. When used in *test(V)*, *sel* and *BLUS* perform equally well, on average, and *z* sometimes inflicts a considerable loss of power.

*Table 5.8* Powers (× 1000) at significance level $\alpha$ of the heterovariance *test (S)* using the six disturbance estimators, applied to the five 15 × 3 **X**-matrices ($\mathcal{H}_A : \eta = 0.9$).

| | $\mathbf{X}_C$ | $\mathbf{X}_K$ | $\mathbf{X}_S$ | $\mathbf{X}_T$ | $\mathbf{X}_A$ | Average |
|---|---|---|---|---|---|---|
| | | | $\alpha = 0.05$ | | | |
| $\hat{u}$ | 407 | 368 | 432 | 439 | 407 | 411 |
| *hyp* | 394 | 349 | 413 | 423 | 412 | 398 |
| *sel* | 394 | 349 | 413 | 423 | 359 | 388 |
| *z* | 327 | 338 | 412 | 273 | 192 | 308 |
| *mod. BLUS* | 347 | 344 | 379 | 392 | 350 | 362 |
| *BLUS* | 341 | 338 | 377 | 386 | 350 | 358 |
| | | | $\alpha = 0.10$ | | | |
| $\hat{u}$ | 536 | 504 | 558 | 567 | 538 | 541 |
| *hyp* | 525 | 486 | 540 | 554 | 542 | 529 |
| *sel* | 525 | 477 | 539 | 554 | 489 | 517 |
| *z* | 445 | 465 | 538 | 403 | 296 | 429 |
| *mod. BLUS* | 479 | 479 | 509 | 523 | 479 | 494 |
| *BLUS* | 473 | 474 | 507 | 516 | 479 | 490 |

The average powers of the tests are pretty small, in particular the powers of *test(V)*. *Test(S)* is *UMPS* with respect to a certain class of **X**-matrices, see Section 2.5, and in view of our experience from *test (Q)*, we believe that the power level cannot be raised substantially, given the specification of $\mathcal{H}_A$. The same argument does not apply to

*Table 5.9* Powers (× 1000) at significance level $\alpha$ of the heterovariance *test(V)* using five disturbance estimators, applied to the five 15 × 3 X-matrices ($\mathcal{H}_A : \gamma = 0.83$).

| | $X_C$ | $X_K$ | $X_S$ | $X_T$ | $X_A$ | Average |
|---|---|---|---|---|---|---|
| | | | $\alpha = 0.05$ | | | |
| $\hat{u}$ | 216 | 237 | 243 | 242 | 233 | 234 |
| *hyp* | 217 | 237 | 230 | 232 | 237 | 231 |
| *sel* | 217 | 229 | 222 | 232 | 218 | 224 |
| *z* | 162 | 207 | 224 | 190 | 140 | 185 |
| *BLUS* | 207 | 232 | 235 | 236 | 219 | 226 |
| | | | $\alpha = 0.10$ | | | |
| $\hat{u}$ | 346 | 370 | 376 | 377 | 366 | 367 |
| *hyp* | 347 | 370 | 361 | 366 | 370 | 363 |
| *sel* | 347 | 358 | 351 | 366 | 345 | 353 |
| *z* | 272 | 333 | 356 | 310 | 241 | 302 |
| *BLUS* | 331 | 362 | 366 | 367 | 346 | 354 |

*test(V)*. Theil (1968) constructed the test statistic $V$ on the basis of a *BLUS* disturbance estimator rather than on the basis of a well-defined alternative hypothesis, thus obtaining an elegant probability distribution at the cost, perhaps, of a loss of power. The difference between the average power levels of *test(S)* and *test(V)* cannot be attributed completely to the difference between the test statistics, since the alternative hypotheses in the two tests differ. For a fair comparison of the two test statistics, it is necessary to consider the powers of two mixed tests, as follows. Let $\mathcal{H}_A(S)$ and $\mathcal{H}_A(V)$ denote the alternative hypotheses in *test(S)* and *test(V)*, respectively. The first mixed test is: use $V$ to test against $\mathcal{H}_A(S)$; and the second mixed test is: use $S$ to test against $\mathcal{H}_A(V)$. Note that both alternative hypotheses let the variances of the disturbances vary from $\sigma^2$ to $10\sigma^2$ when $n = 15$, these variances (apart from $\sigma^2$) being:

$$\mathcal{H}_A(S): \quad 1.06, 1.14, 1.22, \ldots, 1/(1-0.06i), \ldots, 4.55, 6.25, 10.00$$
$$\mathcal{H}_A(V): \quad 1.00, 1.78, 2.49, \ldots, i^{0.83}, \ldots, 8.41, 8.94, 9.47$$

The powers of the two mixed tests, applied to the five 15 × 3 X-matrices, using four disturbance estimators, are presented in Table 5.10.

*Table 5.10* **Powers (x 1000) at significance level 0.05 of the mixed heterovariance tests, using four disturbance estimators, applied to the five 15 x 3 X-matrices.**

| $\mathcal{H}_A$ | Test statistic | Disturbance estimator | $X_C$ | $X_K$ | $X_S$ | $X_T$ | $X_A$ | Average |
|---|---|---|---|---|---|---|---|---|
| $\mathcal{H}_A(S)$ | $V$ | $\hat{u}$ | 294 | 266 | 327 | 334 | 277 | 300 |
| | | *hyp* | 296 | 272 | 318 | 312 | 298 | 299 |
| | | *sel* | 296 | 258 | 287 | 312 | 241 | 279 |
| | | *(mod.)BLUS* | 299 | 294 | 314 | 337 | 283 | 305 |
| $\mathcal{H}_A(V)$ | $S$ | $\hat{u}$ | 235 | 226 | 261 | 256 | 251 | 246 |
| | | *hyp* | 231 | 249 | 249 | 248 | 250 | 245 |
| | | *sel* | 231 | 234 | 248 | 248 | 237 | 240 |
| | | *mod.BLUS* | 202 | 226 | 239 | 234 | 217 | 224 |

Comparing the upper halves of Tables 5.10 and 5.8, we see that, when a test against $\mathcal{H}_A$ $(S)$ is required, the best test statistic is $S$ using *hyp*. Analogously, the upper half of Table 5.9 and the lower half of Table 5.10 reveal that, against $\mathcal{H}_A$ $(V)$, the best test statistic is again $S$ using *hyp*. Therefore, when the problem is to test against gradually increasing heterovariance, we recommend calculating $S$ using *hyp*:

$$S = \frac{1}{n} \sum_{i=1}^{n} iw_i^2 / \sum_{i=1}^{n} w_i^2$$

where $w_i$ is the $i$th element of $\mathbf{w}$, which vector may be calculated in accordance with (4.A.5), taking $\mathbf{P} = [\mathbf{h}_1^* : \mathbf{h}_2^* : \ldots : \mathbf{h}_k^*]$. Then reject $\mathcal{H}_0$ (no heterovariance) if $S \geqslant s$, otherwise accept $\mathcal{H}_0$. Significance points $s$ are presented in Table 5.12. For the purpose of making a balanced choice between the significance level and the power of the test against $\mathcal{H}_A$ $(S)$, we supply Table 5.11.

*Table 5.11* Powers (× 1000) at significance level $\alpha$ of *test(S)* using *hyp*, applied to the $n \times k$ matrix $\mathbf{X} = [\mathbf{h}_1^* : \mathbf{h}_2^* : \ldots : \mathbf{h}_k^*]$ where $\mathcal{H}_A : \eta = 0.90$ and $\eta = 0.95$.

| | | k = 2 | | k = 3 | | k = 4 | | k = 5 | |
|---|---|---|---|---|---|---|---|---|---|
| $n$ | $\alpha$ | $\eta$=0.90 | $\eta$=0.95 | $\eta$=0.90 | $\eta$=0.95 | $\eta$=0.90 | $\eta$=0.95 | $\eta$=0.90 | $\eta$=0.95 |
| 10 | 0.05 | 323 | 448 | 280 | 387 | 239 | 326 | 199 | 264 |
| | 0.10 | 450 | 571 | 404 | 512 | 357 | 450 | 309 | 384 |
| | 0.15 | 537 | 648 | 491 | 594 | 447 | 538 | 395 | 471 |
| | 0.20 | 605 | 707 | 562 | 658 | 519 | 605 | 470 | 544 |
| 15 | 0.05 | 453 | 598 | 418 | 556 | 387 | 515 | 354 | 472 |
| | 0.10 | 580 | 706 | 547 | 671 | 516 | 634 | 485 | 597 |
| | 0.15 | 661 | 771 | 632 | 741 | 604 | 710 | 572 | 674 |
| | 0.20 | 723 | 818 | 695 | 790 | 670 | 764 | 639 | 731 |
| 20 | 0.05 | 555 | 704 | 530 | 676 | 502 | 645 | 480 | 618 |
| | 0.10 | 676 | 796 | 654 | 775 | 630 | 751 | 606 | 726 |
| | 0.15 | 751 | 850 | 731 | 831 | 707 | 810 | 687 | 790 |
| | 0.20 | 802 | 884 | 783 | 868 | 764 | 851 | 745 | 833 |

*Table 5.12* Significance points for *test (S)* using *hyp*.

| | Significance level | | | | Significance level | | | |
|---|---|---|---|---|---|---|---|---|
| $n$ | 0.05 | 0.10 | 0.15 | 0.20 | 0.05 | 0.10 | 0.15 | 0.20 |
| | k = 2 | | | | k = 3 | | | |
| 10 | 0.745 | 0.704 | 0.676 | 0.653 | 0.752 | 0.710 | 0.681 | 0.657 |
| 11 | 0.734 | 0.694 | 0.666 | 0.644 | 0.740 | 0.699 | 0.671 | 0.648 |
| 12 | 0.723 | 0.685 | 0.658 | 0.637 | 0.729 | 0.689 | 0.662 | 0.640 |
| 13 | 0.714 | 0.677 | 0.651 | 0.630 | 0.719 | 0.681 | 0.654 | 0.633 |
| 14 | 0.706 | 0.670 | 0.644 | 0.624 | 0.711 | 0.673 | 0.648 | 0.627 |
| 15 | 0.698 | 0.663 | 0.639 | 0.619 | 0.703 | 0.667 | 0.642 | 0.622 |
| 16 | 0.692 | 0.657 | 0.634 | 0.615 | 0.696 | 0.661 | 0.636 | 0.617 |
| 17 | 0.686 | 0.652 | 0.629 | 0.610 | 0.689 | 0.655 | 0.631 | 0.613 |
| 18 | 0.680 | 0.647 | 0.625 | 0.607 | 0.684 | 0.650 | 0.627 | 0.609 |
| 19 | 0.675 | 0.643 | 0.621 | 0.603 | 0.678 | 0.645 | 0.623 | 0.605 |
| 20 | 0.670 | 0.639 | 0.617 | 0.600 | 0.673 | 0.641 | 0.619 | 0.602 |
| | k = 4 | | | | k = 5 | | | |
| 10 | 0.759 | 0.717 | 0.686 | 0.661 | 0.768 | 0.723 | 0.693 | 0.667 |
| 11 | 0.747 | 0.705 | 0.676 | 0.652 | 0.754 | 0.712 | 0.682 | 0.657 |
| 12 | 0.736 | 0.695 | 0.667 | 0.644 | 0.743 | 0.701 | 0.672 | 0.648 |
| 13 | 0.725 | 0.686 | 0.659 | 0.636 | 0.732 | 0.692 | 0.663 | 0.640 |
| 14 | 0.716 | 0.678 | 0.651 | 0.630 | 0.722 | 0.683 | 0.656 | 0.634 |
| 15 | 0.708 | 0.671 | 0.645 | 0.624 | 0.714 | 0.675 | 0.649 | 0.628 |
| 16 | 0.701 | 0.664 | 0.639 | 0.619 | 0.706 | 0.669 | 0.643 | 0.622 |
| 17 | 0.694 | 0.658 | 0.634 | 0.615 | 0.699 | 0.662 | 0.638 | 0.617 |
| 18 | 0.688 | 0.653 | 0.630 | 0.611 | 0.692 | 0.657 | 0.633 | 0.613 |
| 19 | 0.682 | 0.648 | 0.625 | 0.607 | 0.686 | 0.652 | 0.628 | 0.609 |
| 20 | 0.677 | 0.644 | 0.622 | 0.604 | 0.680 | 0.647 | 0.624 | 0.606 |

# References


Abrahamse, A.P.J. (1970), "A Test on Disturbance Heterovariance in Least-Squares Regression", Report 7019 of the Econometric Institute, Erasmus University, Rotterdam.

Abrahamse, A.P.J. and J. Koerts (1971), "New Estimators of Disturbances in Regression Analysis", *Journal of the American Statistical Association*, 66, pp. 71-74.

Abrahamse, A.P.J. and A.S. Louter (1971), "On a New Test for Autocorrelation in Least-Squares Regression", *Biometrika*, 58, pp. 53-60.

Abramowitz, M. and I.A. Stegun (1965), *The Handbook of Mathematical Functions*, New York.

Anderson, T.W. (1948), "On the Theory of Testing Serial Correlation", *Skandinavisk Aktuarietidskrift*, XXXI, pp. 88-116.

Anderson, T.W. (1971), *The Statistical Analysis of Time Series*, New York.

Berenblut, I.I. and G.I. Webb (1973), "A New Test for Autocorrelated Errors in the Linear Regression Model", *Journal of the Royal Statistical Society*, B 35, pp. 33-50.

Chow, G.C. (1957), *Demand for Automobiles in the United States*, Amsterdam.

Dubbelman, C., A.P.J. Abrahamse, and A.S. Louter (1976), "On Typical Characteristics of Economic Time Series and the Relative Qualities of Five Autocorrelation Tests", Report 7604 of the Econometric Institute, Erasmus University, Rotterdam.

Durbin, J. (1970), "An Alternative to the Bounds Test for Testing for Serial Correlation in Least-Squares Regression", *Econometrica*, 38, pp. 422-429

Durbin, J. and G.S. Watson (1950), "Testing for Serial Correlation in Least-Squares Regression. I", *Biometrika*, 37, pp. 409-428.

Durbin, J. and G.S. Watson (1951), "Testing for Serial Correlation in Least-Squares Regression. II", *Biometrika*, 38, pp. 159-178.

Durbin, J. and G.S. Watson (1971), "Testing for Serial Correlation in Least-Squares Regression. III", *Biometrika*, 58, pp. 1-19.

Geary, R.C. (1966), "A Note on Residual Heterovariance and Estimation Efficiency in Regression", *The American Statistician*, 20 Nr. 4, pp. 30-31.

Hannan, E.J. (1960), *Time Series Analysis*, London.

Henshaw, R.C. (1966), "Testing Single-Equation Least-Squares Regression Models for Autocorrelated Disturbances", *Econometrica*, 34, pp. 646-660.

Hildebrand, F.B. (1956), *Introduction to Numerical Analysis*, New York.

Imhof, P.J. (1961), "Computing the Distribution of Quadratic Forms in Normal Variables", *Biometrika*, 48, pp. 419-426.

Klein, L.R. (1950), *Economic Fluctuations in the United States 1921-1941*, New York.

Kmenta, J. (1971), *Elements of Econometrics*, London.

Koerts, J. (1965), *Schattingsfuncties van Storingen in Economische Relaties*, Rotterdam.

Koerts, J. and A.P.J. Abrahamse (1969), *On the Theory and Application of the General Linear Model*, Rotterdam.

Lehmann, E.L. (1959), *Testing Statistical Hypotheses*, New York.

L'Esperance, W.L., D. Chall, and D. Taylor (1976), "An Algorithm for Determining the Distribution Function of the Durbin-Watson Test Statistic", *Econometrica*, 44, pp. 1325-1326.

Louter, A.S. and C. Dubbelman (1973), "An Exact Autocorrelation Test for Small n and k, a Computer Program and a Table of Significance Points", Report 7304 of the Econometric Institute, Erasmus University, Rotterdam.

Malinvaud, E. (1970), *Statistical Methods of Econometrics*, Amsterdam
Pan Jie-Jian (1968), "Distribution of the Noncircular Serial Correlation Coefficients", Am. Math. Soc. and Inst. Math. Statist. *Selected Translations in Probability and Statistics*, 7, pp. 281-291.
Ramsey, J.B. (1969), "Tests for Specification Errors in Classical Linear Least-Squares Regression Analysis", *Journal of the Royal Statistical Society*, 31, pp. 350-371.
Sato, R. (1970), "The Estimation of Biased Technical Progress and the Production Function", *International Economic Review*, 11, pp. 179-208.
Sims, C.A. (1975), "A Note on Exact Tests for Serial Correlation", *Journal of the American Statistical Association*, 70, pp. 162-165.
Theil, H. (1965), "The Analysis of Disturbances in Regression Analysis", *Journal of the American Statistical Association*, 60, pp. 1067-1078.
Theil, H. (1968), "A Simplification of the BLUS Procedure for Analysing Regression Disturbances", *Journal of the American Statistical Association*, 63, pp. 242-251.
Theil, H. (1971), *Principles of Econometrics*, Amsterdam.
Von Neumann, J. (1941), "Distribution of the Mean Square Successive Difference to the Variance", *Annals of Mathematical Statistics*, 12, pp. 367-395.

# Index